# Reviews of Environmental Contamination and Toxicology

VOLUME 196

# Reviews of Environmental Contamination and Toxicology

Editor
David M. Whitacre

Editorial Board
Lilia A. Albert, Xalapa, Veracruz, Mexico • Charles P. Gerba, Tucson, Arizona, USA
John Giesy, Saskatoon, Saskatchewan, Canada • O. Hutzinger, Bayreuth, Germany
James B. Knaak, Getzville, New York, USA
James T. Stevens, Winston-Salem, North Carolina, USA
Ronald S. Tjeerdema, Davis, California, USA • Pim de Voogt, Amsterdam, The Netherlands
George W. Ware, Tucson, Arizona, USA

Founding Editor
Francis A. Gunther

VOLUME 196

# Coordinating Board of Editors

Dr. David M. Whitacre, *Editor*
*Reviews of Environmental Contamination and Toxicology*

5115 Bunch Road
Summerfield North, Carolina 27358, USA
(336) 634-2131 (PHONE and FAX)
E-mail: dmwhitacre@triad.rr.com

Dr. Herbert N. Nigg, *Editor*
*Bulletin of Environmental Contamination and Toxicology*

University of Florida
700 Experiment Station Road
Lake Alfred, Florida 33850, USA
(863) 956-1151; FAX (941) 956-4631
E-mail: hnn@LAL.UFL.edu

Dr. Daniel R. Doerge, *Editor*
*Archives of Environmental Contamination and Toxicology*

7719 12th Street
Paron, Arkansas 72122, USA
(501) 821-1147; FAX (501) 821-1146
E-mail: AECT_editor@earthlink.net

ISBN: 978-1-4419-2688-3        e-ISBN: 978-0-387-78444-1
DOI: 10.1007/978-0-387-78444-1

© 2010 Springer Science+Business Media, LLC
All rights reserved. This work may not be translated or copied in whole or in part without the written permission of the publisher (Springer Science+Business Media, LLC, 233 Spring Street, New York, NY 10013, USA), except for brief excerpts in connection with reviews or scholarly analysis. Use in connection with any form of information storage and retrieval, electronic adaptation, computer software, or by similar or dissimilar methodology now known or hereafter developed is forbidden.
The use in this publication of trade names, trademarks, service marks, and similar terms, even if they are not identified as such, is not to be taken as an expression of opinion as to whether or not they are subject to proprietary rights.
While the advice and information in this book are believed to be true and accurate at the date of going to press, neither the authors nor the editors nor the publisher can accept any legal responsibility for any errors or omissions that may be made. The publisher makes no warranty, express or implied, with respect to the material contained herein.

Printed on acid-free paper

9 8 7 6 5 4 3 2 1

springer.com

# Contents

**Environmental Impact of Pesticides in Egypt** ............................................. 1
Sameeh A. Mansour

**Biodegradation of Perfluorinated Compounds** .......................................... 53
John R. Parsons, Monica Saéz, Jan Dolfing, and Pim de Voogt

**Lead Stress Effects on Physiobiochemical Activities
of Higher Plants** ............................................................................................ 73
Rakesh Singh Sengar, Madhu Gautam, Rajesh Singh Sengar,
Sanjay Kumar Garg, Kalpana Sengar, and Reshu Chaudhary

**Environmental Fate and Toxicology of Carbaryl** ....................................... 95
Amrith S. Gunasekara, Andrew L. Rubin, Kean S.Goh,
Frank C. Spurlock, and Ronald S. Tjeerdema

**Managing Hazardous Pollutants in Chile: Arsenic** .................................... 123
Ana María Sancha and Raul O'Ryan

**Atrazine Interaction with Estrogen Expression Systems** ........................... 147
J. Charles Eldridge, James T. Stevens, and Charles B. Breckenridge

**Index** ............................................................................................................. 161

# Environmental Impact of Pesticides in Egypt

Sameeh A. Mansour

**Contents**

| | | |
|---|---|---|
| 1 | Introduction................................................................................................ | 1 |
| 2 | The Egyptian Pesticide Market.................................................................. | 2 |
| 3 | Environmental Impact of Pesticides ......................................................... | 5 |
| 4 | Problems Associated with Pesticides in Egypt......................................... | 6 |
| | 4.1 Incidental Toxicity to Humans......................................................... | 6 |
| | 4.2 Incidental Toxicity to Farm Animals .............................................. | 10 |
| | 4.3 Pest Resistance................................................................................. | 11 |
| | 4.4 Methyl Parathion Spill at Port-Said, Egypt.................................... | 11 |
| | 4.5 Accumulation of Pesticide Residues in Natural Ecosystems........ | 12 |
| | 4.6 Food Contamination ........................................................................ | 22 |
| | 4.7 Pesticide Residues in Human Blood and Urine............................. | 33 |
| | 4.8 Risks of Dietary Pesticide Residues ............................................... | 34 |
| 5 | Factors Contributing to Pesticide Hazards: Humans and the Environment...... | 40 |
| | 5.1 General............................................................................................... | 40 |
| | 5.2 Specific .............................................................................................. | 41 |
| 6 | Conclusions................................................................................................ | 44 |
| 7 | Summary..................................................................................................... | 46 |
| References | | 47 |

## 1 Introduction

Egypt is the most populous country in the Arab world, and the second most populous in Africa after Nigeria. Egyptians live on 3 million ha, and this area must feed 80 million citizens. Modest calculations indicate that by 2018 the country's population may exceed 100 million (Mansour 2004). Agricultural activities account for 28% of total national income, and nearly half of the country's work force is dependent on the agricultural subsector for its livelihood. Moreover, Egypt is rapidly industrializing

---

S.A. Mansour
Professor of Pesticides and Environmental Toxicology,
Pesticide Chemistry Department, National Research Centre
Tahir Str., Dokki Cairo, Egypt
e-mail: samansour@hotmail.com

and is using chemicals extensively in a wide spectrum of industrial sectors. Some 4 million metric tons (t) of chemicals and chemical products are annually imported into Egypt. These imported chemicals represent about 95% of manufactured chemicals found and used in the country. The major industrial sectors in Egypt are raw and fabricated metals, vehicles, and pharmaceuticals. Other important industries include textiles, pesticides, fertilizers, petrochemicals, cement, paper and pulp, and food processing. About 50% of all industrial activity is concentrated in greater Cairo (Cairo, Giza, and Qalubiya) and about 40% in Alexandria. The remaining industrial activity rests in the Delta Region and Upper Egypt and in newer cities.

Egypt is an arid country, composed of 97% desert by area, and is dependent on the Nile River for its existence. Because of the intensive irrigation needed to support the desert landscape, Egypt has developed a host of environmentally related problems. These problems have been aggravated by high population density, which places a further strain upon resources. In addition, modern development has imposed great stress on Egypt's environment. The chemical industry in Egypt is, by far, the main source of hazardous waste release in developed regions. These industries have encountered frequent problems in disposing of the hazardous waste they generate. In addition to the foregoing pollution, water pollution is exacerbated by agricultural pesticides, raw sewage, and urban and industrial effluents (Barakat 2004).

According to a recent report ("Poverty and Poverty Alleviation Strategies in Egypt") to the Ford Foundation (Assaad and Roushdy 1998), 25% of Egypt's population is very poor and another 25% lives near the poverty line. Another source suggests that almost one-third of Egypt's population lives below the poverty line, defined as a total income less than $U.S. 30.00 per month (UNDP, Egypt 1999). Thus, Egypt has the challenge of maximizing its limited resources to deal with an array of problems. The tension between population increase and production of sufficient food for all is one of the most important and challenging problems currently facing Egypt. Therefore, agricultural productivity is crucial to Egypt, and the use of pest control agents (e.g., pesticides) to enhance productivity will remain as an essential contributing component for the foreseeable future. At the global level, the use of pesticides has assisted in solving many human health and food production problems. However, such usage has also frequently been accompanied by hazards to humans and the environment.

The objective of this review is to provide a comprehensive assessment of the threats posed by pesticide contamination to human health and the environment in Egypt. In addition, it endeavors to identify priorities and data gaps and to make recommendations for appropriate future interventions adequate to control or reduce pesticide contamination.

## 2 The Egyptian Pesticide Market

Cotton still represents the most important crop and a main element in Egypt's national economy. Pests infesting cotton affect the crop's quality and yield. Pesticides are considered to be one of the major elements in protecting cotton production. Insecticides are applied to an area of 0.28–0.32 million ha by high-pressure ground sprayers. Applications are made to this area annually three to five times per season to control the cotton leafworm (*Spodoptera littoralis*), the pink bollworm (*Pectinophera gossypiella*), the spiny bollworm (*Erias insulana*), and many other pests. Other crops are treated with pesticides as well. Corn, rice, sugar cane, and many different varieties of vegetables and fruits consume large quantities of various pesticides (insecticides, fungicides, herbicides, etc.).

Before 1950, only limited Egyptian cotton acreage was treated with insecticides. Thereafter, the treated area expanded rapidly. During the period 1950–1955, scattered cotton fields were treated using insecticidal dusts (10% DDT + 25% BHC + 40% sulfur) (El-Sebae and Soliman 1982). Table 1 presents the types and amounts of insecticides used on cotton from 1952 to 1990 (El-Sebae et al. 1993). As shown by these data, toxaphene was the most abundantly used insecticide from 1955 to 1961. The continuous shifting from one insecticide to another was mainly attributed to the development of resistance by the cotton leafworm.

Mansour (1993, 2004) published data showing the average annual consumption of pesticides during the 1970s, 1980s, and 1990s, by decade; production totals were 26,029, 24,369, and 17,000 t, respectively. Recently (2000–2006), consumption of pesticides has gradually declined, reaching about 12,000 t annually (Table 2). The local pesticide market obtains the pesticides it needs through two main sources: public and private. The Ministry of Finance (public source) imports pesticides to meet needs set by the Ministry of Agriculture (MOA). The private sector finances pesticide importation to meet its marketing or use needs using its own resources.

**Table 1** Total active ingredient (a.i.s.) insecticidal use on cotton in Egyptian agriculture from 1952 to 1990

| Compound | Total metric tons (t) | Year of consumption |
|---|---|---|
| Toxaphene | 54,000 | 1955–1961 |
| Endrin | 10,500 | 1961–1981 |
| DDT | 13,500 | 1952–1971 |
| Lindane | 11,300 | 1952–1978 |
| Carbaryl | 21,000 | 1961–1978 |
| Trichlorfon | 6,500 | 1961–1970 |
| Monocrotophos | 8,300 | 1967–1978 |
| Leptophos | 5,500 | 1968–1978 |
| Chlorpyrifos | 13,500 | 1969–1990 |
| Phosfolan | 5,500 | 1963–1983 |
| Mephosfolan | 7,000 | 1968–1983 |
| Methamidophos/azinphos-methyl | 7,500 | 1970–1990 |
| Triazophos | 8,500 | 1977–1990 |
| Profenofos | 8,000 | 1977–1990 |
| Methomyl | 9,500 | 1975–1990 |
| Fenvalerate | 8,500 | 1976–1990 |
| Cypermethrin | 6,300 | 1976–1990 |
| Deltamethrin | 5,400 | 1976–1990 |
| Cyanophos | 3,000 | 1984–1990 |
| Thiodicarb | 5,000 | 1984–1990 |

*Source*: El-Sebae et al. (1993).

**Table 2** Average annual consumption of pesticides in Egypt by decade from 1970 to 2006

| Parameter | Metric tons (t) |
|---|---|
| 1970–1979[a] | 26,029 |
| 1980–1989[a] | 24,369 |
| 1990–1999[b] | 17,000 |
| 2000–2006[c] | 12,000 |

[a]*Source*: Mansour (1993).
[b]*Source*: Mansour (2004).
[c]Author's estimate.

**Table 3** Number of active ingredients (a.i.s) by pesticide class used in Egypt from 1980 to 2002

| Year (growing season) | Number of a.i.s | | | | | | |
|---|---|---|---|---|---|---|---|
| | Insecticides | Fungicides | Herbicides | Acaricides | Nematicides | Rodenticides | Total |
| 1980/1981 | 71 | 43 | 36 | 7 | 5 | 4 | 166 |
| 1990/1991 | 71 | 69 | 53 | 8 | 5 | 15 | 221 |
| 2001/2002 | 59 | 60 | 36 | 10 | 4 | 6 | 175 |

*Source*: Official Recommendations Books issued by the Egyptian Ministry of Agriculture.

By the end of 1989, the private sector accounted for 26% of total pesticides consumed in Egypt (total imported, US $100 million a-nnually). Pesticides formulated locally at that time represented 48.5% of the total quantity consumed in the country (Mansour 1993). In the past decade, local formulating plants have greatly increased their share of the market, as has the private sector.

The distribution pattern by class of pesticides used in 1990 was as follows: insecticides (55%), fungicides (16.0%), rodenticides (15.0%), and herbicides (13.0%) (Mansour 1993). Table 3 shows that a total of 166 active ingredient (a.i.s.) pesticides were recommended for use in Egypt during the 1980–1981 growing season. Of this number 71, 43, and 36 a.i.s. comprised insecticides, fungicides, and herbicides, respectively. The 1990–1991 growing season saw a jump in the total number of pesticides (221). This increase was mainly attributed to new recommendations for fungicides and herbicides. During 2001–2002, the number of pesticides registered in Egypt reached 330 formulations, representing 175 a.i.s. Insecticides and fungicides each represent 34% of the total number of a.i.s. used in Egypt. Herbicides, acaricides, rodenticides, and nematicides represent 21%, 6%, 3%, and 2% of the total, respectively. Included in these classes are compounds belonging to the World Health Organization (WHO) Class IA, "extremely hazardous" pesticides (e.g., aldicarb, chlorophacinone, difenacoum, diphacinone), as well as Class IB "highly hazardous" compounds (e.g., carbosulfan, fenamiphos, methomyl, oxamyl, triaziphos). Moreover, pesticides such as propargite, mancozeb, maneb, folpet, procymidone, captan, and cyproconazole of "Group B," as well as dimethoate, carbaryl, ethofenprox, dicofol, iprodione, benomyl, triadimefon, atrazine, oxyfluorfen, oxadiazon, linuron, simazine, and pendimethalin of "Group C," were available for use in local formulations (Mansour 2004). Since mid-2003, the MOA undertook certain actions in an attempt to replace these dangerous pesticides with safer alternatives. In 2005, the Ministerial Decree No. 719 prohibited the use, importation, and local formulation of 47 pesticide a.i.s. However, several of these (e.g., captan, chlorothalonil, cypermethrin, folpet, fosetyl aluminum, iprodione, mancozeb, maneb, propiconazole, thiabendazole, and thiophanate-methyl) have recently applied for official re-use in the country under another Ministerial Decree (No. 630) for the year 2007. The use of certain pesticides, such as dicofol, dimethoate, etofenprox, oxadiazon, oxyfluorfen, procymidone, propargite, tebuconazole, tetraconazole, triadimenol, and trifluralin (which were included in the Ministerial Decree No. 719), will be reviewed in light of European Union (EU) recommendations. On the other hand, the last Decree (No. 630, dated 3 May 2007) listed a total of 371 pesticide a.i.s. as completely banned in Egypt.

The use of pesticides in Egypt is governed by two laws: Act No. 53 (1966) issued by the Minister of Agriculture for agricultural use pesticides, and a separate law (Act No. 127, 1955) for household pesticides issued by the Minister of Health. Subsequent amendments were added to both laws according to need. The main provisions of the Agricultural Pesticides Act and its regulations were previously reviewed by many authors (e.g., Mansour 1993). Briefly, the registration scheme for pesticides in this Act is consonant with the Food and Agriculture Organization of the United Nations (FAO) Guidelines on the Registration and Control of Pesticides.

## 3 Environmental Impact of Pesticides

As previously mentioned, a huge quantity of organochlorine pesticides (OCP) was used in Egypt between 1950 and 1981 to protect crops from insects, disease, and weeds, to remove unwanted vegetation, and to control indoor insects to which the general public was exposed. Over time, these persistent compounds were replaced by other chemical classes of shorter persistence.

Pesticides move through air, soil, and water and find their way into living tissues where they can bioaccumulate through the food chain, eventually to enter the human diet. Approximately 85%–90% of applied agricultural pesticides never reach target organisms, but disperse through the air, soil, and water (Moses et al. 1993). Persistent pesticides can remain for decades; the half-life of toxaphene in soil, for example, is up to 29 yr (PAN 1993). Pesticides that are not bound in soils or taken up into plants and animals can run off into rivers and lakes and move into the aquatic food chain, inducing severe damage to aquatic life. Such environmental mobility can cause contamination of several environmental compartments. Pesticides, thereby, augment other sources of environmental pollution in Egypt, which include manufacturing processes such as the pulp and paper industry, textile and leather dying, and thermal processes in the metallurgical, cement, motor vehicle, and steel industries. Petroleum hydrocarbons are potentially the most likely source of polyaromatic hydrocarbon (PAHs) pollutants. Other important potential sources of PAHs that do not arise from burning fossil fuel come from combustion of domestic wastes in largely uncontrolled situations, which although intermittent in nature can have a very marked effect on air quality (Barakat 2003).

## 4 Problems Associated with Pesticides in Egypt

### *4.1 Incidental Toxicity to Humans*

Pesticides, although widely used throughout the world, are toxic chemicals. There are approximately 600 pesticidal a.i.s. in use worldwide, and a far greater number of formulated commercial products. Pesticides are primarily used in agriculture, although use in public health programs for malaria prevention or rodent control is also significant in some areas of the world (Garcia 1998).

Human exposure to pesticides is high in Egypt, particularly from ground application of rather toxic insecticides to cotton fields. Large numbers of workers, laborers, and overseers are involved in pesticide applications to cotton fields three to five times each season. Pesticides are typically applied to cotton from May to September (approximately 120 d each year), and about 12,000 workers countrywide participate in this task. A typical workday is about 8 hr, and the typical climatic conditions are severe (high temperature and humidity). Safety practices are generally inadequate, and typically workers lack proper training in safe handling of these chemicals (Amr and Halim 1997). Moreover, workers are often not equipped with protective clothing or face masks. In addition, thousands of children, who collect egg masses of the cotton leafworm daily for about 40 d each season, are exposed to insecticide residues on cotton leaves. There are also issues with the large numbers of workers involved in formulating pesticides and handling them in greenhouses. The foregoing will appropriately give rise to the consideration that humans in Egypt have a relatively high exposure to pesticides.

Official reports of pesticide poisoning in Egypt are lacking, except for certain dramatic cases released to the public and a few reports published by scientists. The only available data were published by El-Gamal (1983), who reported cases of poisoning and numbers of fatalities annually from 1966 to 1982 (Table 4). This author stated that after the year 1977 incidents of acute intoxication decreased as a result of the introduction of exposure prevention measures. El-Gamal also states that more than 60% of workers engaged in pesticide applications suffer from chronic toxicity.

A study was conducted to evaluate the impact on health of workers exposed to pesticides in large- and small-scale Egyptian formulation plants (Amr 1990). Dermatitis and neuropsychiatry manifestations were the most prevalent health effects in this exposed population, compared to the controls, especially on workers with a longer duration of employment. Other manifestations of exposure in this population included topical eye changes and gastrointestinal and genitourinary

Table 4  Incidence of pesticide poisoning in Egypt: 1966–1982

| Year | No. of cases | No. of deaths |
|---|---|---|
| 1966 | 1,091 | 44 |
| 1967 | 1,270 | 34 |
| 1968 | 1,608 | 35 |
| 1969 | 1,389 | 44 |
| 1970 | 1,473 | 67 |
| 1971 | 741 | 21 |
| 1972 | 1,309 | 44 |
| 1973 | 493 | 2 |
| 1974 | 1,951 | 33 |
| 1975 | 967 | 4 |
| 1976 | 510 | 24 |
| 1977 | 2,671 | 69 |
| 1978 | 1,439 | 29 |
| 1979 | 1,062 | 31 |
| 1980 | 569 | 38 |
| 1981 | 491 | 10 |
| 1982 | 1,066 | 42 |

Source: El-Gamal (1983).

effects, as well as hepatomegaly and pulmonary function changes. A significantly higher frequency of polyneuropathy, sensory hypothesia, and abnormal deep reflexes was also observed among exposed workers. The levels of serum gonadotropins [luteinizing hormone (LH) and follicle-stimulating hormone (FSH)] and testosterone were significantly higher in exposed than in control groups, particularly for LH.

In a study at Kafr El-Sheikh Governorate, Egypt, involving 240 individuals, pesticide applicators (PA) had greater reduction of semen quality compared with non-farm workers (NFW). Also, biochemical markers [e.g., uric acid, urea, creatinine, and aspartate aminotransferase (AST)] in PA were near the upper limits of the normal range (Attia 2005).

Ezzat et al. (2005) studied the association between pesticide exposure and hepatocellular carcinoma (HCC) among 236 subjects from urban and rural regions (Table 5). The authors also obtained information on rates of exposure to pesticides in the home or in agricultural fields. Of the total population, 62% were exposed to pesticides at either urban or rural homes. Exposure to rodenticides (in home and field), and other pesticides in agricultural fields was higher among the rural group (Table 5). In rural

Table 5  Proportion (%) of persons exposed to pesticides at home and during agricultural work

| Item | Urban ($n = 68$) (M = 36; F = 32) | Rural ($n = 150$) (M = 113; F = 37) |
|---|---|---|
| Pesticides at home | 53 (61.6%) | 93 (62.0%) |
| Rodenticides at home | 7 (8.1%) | 22 (14.7%) |
| Pesticides, field work | 5 (5.8%) | 81 (54.0%) |
| Rodenticides, field work | 4 (4.7%) | 55 (36.7%) |

$n$, Total number of cases; M, males; F, females.
Source: Adapted from Ezzat et al. (2005).

males ($n$ = 113), 54.9% were exposed to carbamate pesticides and 63.7% were exposed to organophosphorus (OP) compounds. Almost one-third of workers (29.2%) were exposed to dithiocarbamate fungicides. These data clearly reveal the multidimensional nature of worker exposure to different pesticide classes.

In a field study, farm workers were occupationally exposed to OP pesticides through their application to cotton fields in Menoufiya Governorate, Egypt (Farahat et al. 2007). Results indicated that exposed individuals exhibited significantly lower performance than did nonexposed controls on six neurological tests (similarities, digital symbol, trail making part A and B, letter cancellation, digital span, and Benton visual retention). Serum acetyl cholinesterase (AChE) was significantly lower in exposed (87.34 U/mL) than control (108.25 U/mL) participants, and longer duration of work with pesticides was associated with a lower AChE level.

## Poison Control Center Statistics

Alexandria Poison Center (APC)

Abdelmegid and Salem (1996) surveyed 5,913 patients admitted to the APC during 1994. Patients 15–35 yr old represented 52.3% of admissions, followed by those less than 5 yr of age (19.4%). Slightly less than one-quarter of patients (24.7%) suffered food poisoning. Poisoning by household agents (e.g., Clorox, kerosene, potash, phenol, tanning chemicals, benzene, sulfuric acid, and antiseptics) constituted 21.2% of admissions, followed by those poisoned with drugs and pesticides (18.2% and 14.3%, respectively). More females (16%) than males (13%) were poisoned by pesticides; most of those poisoned were between the ages 15 and 25 yr. Generally, the highest number of poisoning cases was recorded during the months of July–September (Table 6).

## Poison Control Center of Ain Shams University (PCCA) Hospitals[1]

The PCCA is the largest poisoning center in Egypt. The total number of acute poisoning cases received at PCCA from 1982 to 2006 is presented in Fig. 1. During the last 4 yr, the number of those poisoned ranged between 21,805 and 25,555 cases/yr. In 2006 (21,805 cases), the great majority of poisonings (94.0%) occurred via the oral route. Non-drug substances accounted for 54.0% of poisonings (11,614 cases), while drugs represented 37.0%. Poisoning by non-drugs included chemicals (6,952 cases; 60.0%), food poisoning (28.0%), animal poisoning (venom of scorpion and snakes) (6.0%), and other (Fig. 2). The main chemicals causing poisoning were insecticides, corrosives, petroleum distillates, and other chemicals, representing 51.0%, 22.0%, 13.0%, and 14.0%, respectively.

---

[1] *These data were kindly provided by the PCCA's Director.*

**Table 6** Pattern of acute human poisonings in 1994 from records of the Alexandria Poison Center (APC)

| Type of poisoning | Male | | Female | | Total | |
|---|---|---|---|---|---|---|
|  | No. | % | No. | % | No. | % |
| Food | 815 | 24.3 | 640 | 25.0 | 1,455 | 24.7 |
| Household agents | 748 | 22.3 | 507 | 19.8 | 1,255 | 21.2 |
| Drugs | 371 | 11.1 | 702 | 27.5 | 1,073 | 18.1 |
| Pesticides | 436 | 13.0 | 410 | 16.0 | 846 | 14.3 |
| Alcohol | 500 | 14.9 | 16 | 0.6 | 516 | 8.7 |
| Gases | 103 | 3.1 | 141 | 5.5 | 244 | 4.1 |
| Addiction | 217 | 6.4 | 9 | 0.4 | 226 | 3.8 |
| Unknown | 79 | 2.3 | 103 | 4.0 | 182 | 3.1 |
| Animal poisons | 52 | 1.5 | 27 | 1.1 | 79 | 1.4 |
| Plant poisons | 36 | 1.1 | 1 | 0.1 | 37 | 0.6 |
| Totals | 3,357 | 100.0 | 2,556 | 100.0 | 5,913 | 100.0 |

Total number of patient poisonings, 5913; sex ratio, 1.3:1.0 (male:female); age of patients: <5 yr, 19.4%; 5–10 yr, 14.3%; 15–35 yr, 52.3%; >45 yr, 5.7%.
*Source*: Adapted from Abdelmegid and Salem (1996).

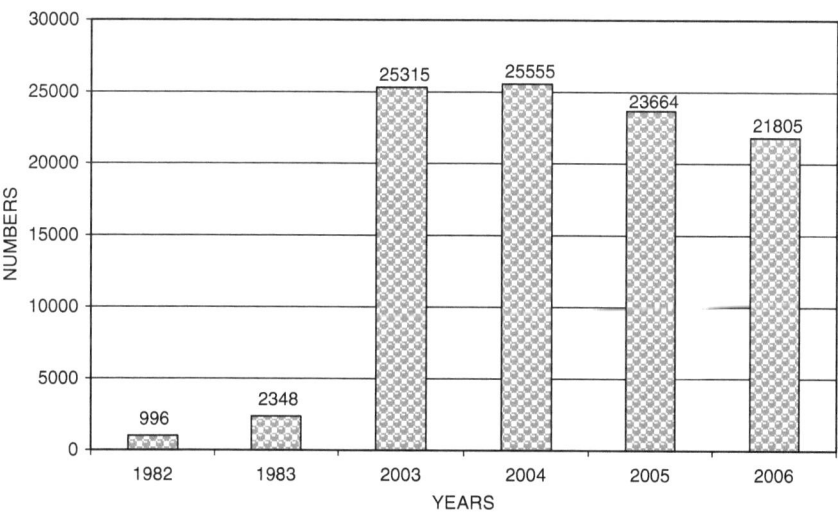

**Fig. 1** Total number/year of poisoning patients received at the Poison Control Center of Ain Shams University Hospitals (PCCA). (Personal communication from the Director of PCCA.)

The number of insecticide poisoning cases in 2006 reached 3,564. Of this number, OP insecticides accounted for 75.0%, while carbamates (5.0%) and zinc phosphide (20.0%) accounted for fewer poisoning incidents.

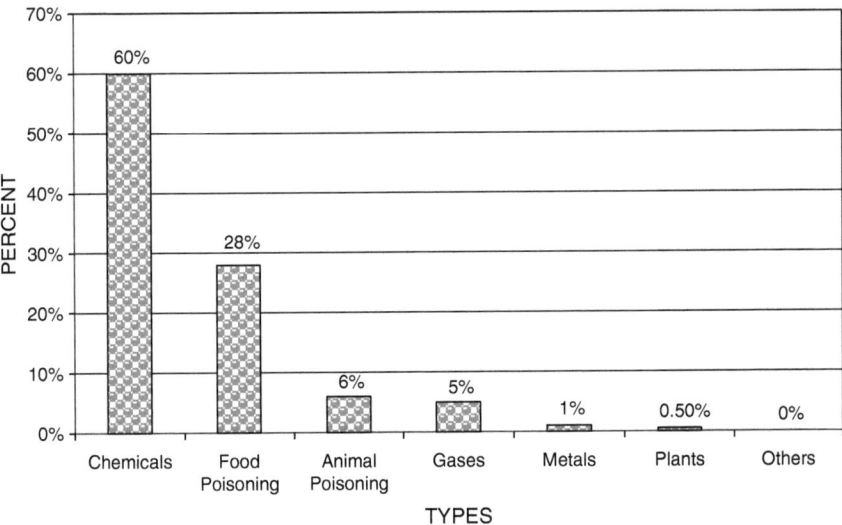

**Fig. 2** Types of non-drug poisoning among patients admitted at PCCA in 2006

## 4.2 Incidental Toxicity to Farm Animals

The failure of several insecticides to overcome cotton leafworm resistance led to the introduction of newer compounds that were not yet registered in the producing countries (e.g., phosfolan, mephosfolan, leptophos). As a result of the introduction of these new chemicals, El-Sebae (1977) reported incidents of delayed neurotoxicity to farm animals from large-scale applications of leptophos (Phosvel®) to cotton during the period 1971–1974. In 1971, a mysterious epidemic of paralysis struck several hundred water buffalo at villages in the middle Delta region, eventually resulting in the death of 1,300 animals. Evidence strongly pointed to leptophos as being the agent responsible for this delayed neurotoxic syndrome. Additional similar incidents were recorded in 1971 in Dakahliya and Sharkiya Provinces. In 1973 and 1974, other incidents were recorded in Fayoum and Minya Provinces, respectively, because of to spraying of leptophos or EPN (an OP insecticide, acaricides) blends. Furthermore, some people developed neurotoxic poisoning symptoms, and traces of leptophos were found in their tissues. Other OP insecticides (e.g., trichloronate, salithion, cyanophos, methamidophos, trichlorfon, DDVP) were found to cause delayed neuropathy to some degree in humans and animals (El-Sebae et al. 1979, 1981).

## 4.3 Pest Resistance

In Egypt, it is well known that most insecticide use is directed to cotton fields. Therefore, the side effects most often exerted by insecticides result from applications on cotton.

The use of an insecticide against a particular pest may cause (a) development of resistance in the pest population, (b) destruction of natural pest enemies (e.g., predators and parasites), and (c) appearance of new pests.

It was first proposed that toxaphene (60% Emulsifiable Concentrate (EC) be introduced to Egypt in 1955 to combat major cotton insect pests (e.g., *Spodoptera littoralis*, *Pectinophera gossypiella*, *Erias insulana*). During the 1956–1958 seasons, the insecticide was used in four successive sprays per season at a rate of 2 L/ac (1200 g a.i.s./A). The rate was then increased to 3 L/A during the next two seasons (1959–1960). In the 1961 season, a disaster occurred when toxaphene, even at 4 L/ac, failed to control the outbreak of the cotton leafworm, resulting in loss of 50% of the national cotton yield for that season. This failure was caused by the buildup of resistance to this insecticide. The total amount of toxaphene a.i.s. distributed during its six seasons of use was estimated to be 54,000 t.

The $LD_{50}$ of toxaphene to fourth larval instars of *S. littoralis* collected from cotton fields was 11 mg/g compared with 0.42 mg/g for the laboratory strain of this instar; this indicates a 26-fold resistance level of the field population. Field trials showed that cross-resistance to other insecticides had developed, but there was no reversal of toxaphene resistance after it was no longer used. After the toxaphene disaster, many insect resistance problems occurred with other insecticides. Table 7 presents the list of insecticides to which the cotton leafworm *S. littoralis* developed resistance from 1961 to 1972 (El-Sebae et al. 1993).

## 4.4 Methyl Parathion Spill at Port-Said, Egypt

In February 1982, the collision of two ships, namely, *Garnet* and *Molaventure*, took place near the northern entrance to the Suez Canal at Port-Said. As a result, *Garnet*,

**Table 7** Insecticides to which the cotton leafworm, *Spodoptera littoralis*, developed resistance

| Insecticide | Year resistance developed |
|---|---|
| Toxaphene | 1961 |
| DDT | 1963 |
| Lindane | 1964 |
| Endrin | 1965 |
| Carbaryl | 1965 |
| Azinphos-methyl | 1968 |
| Monocrotophos | 1972 |
| Methamidophos | 1972 |

*Source*: El-Sebae et al. (1993).

which was loaded with 31,000 kg methyl parathion, slowly sank in shallow water. More than 10,000 kg of the insecticide found its way into the Mediterranean Sea.

Estimated concentrations of methyl parathion (mid-June 1982) near the area where the ship sank ranged from 1.0 to 96.0 µg/L in water, 5.1 to 450.0 µg/kg in sediment, and 46.7 to 195.5 µg/kg in fish. Bioaccumulation in fish species followed the order *Anguilla > Mugil > Sardine > Scidena* (Badawy et al. 1984).

## 4.5 Accumulation of Pesticide Residues in Natural Ecosystems

**Contamination of Water**

Studies on freshwater aquatic environments have mainly been focused on the Nile River and the major delta lakes of Egypt. The data presented in Table 8 show the persistence of OCP residues (e.g., DDT and its metabolites; HCH and its isomers and cyclodiene compounds) in water of the River Nile and some lakes. This group of insecticides was prohibited from use in Egypt as early as 1982, although residues of the OCPs are still detectable at low concentration levels.

Water of the Damietta estuaries seemed to be more contaminated than that of Rosetta estuaries, based on samples collected for analyses in 1995–1997. The range

**Table 8** Organochlorine (OC) residues in water samples and year(s) collected from different locations in Egypt

| Location | Concentration (ppb) | Location | Concentration (ppb) |
|---|---|---|---|
| Lake Manzala | 51.0–160.2 (1991)[5] | River Nile: | |
|  | 0.005–0.037 (1992)[6] | | |
|  | 35.4–69.4 (1992–1993)[7] | Cairo | 0.015 (1982)[3] |
|  | 0.50–2.3.1 (1993)[8] | | 17.9–72.9 (1991)[5] |
|  | 0.03–0.42 (1997–1998)[13] | | 0.13–0.19 (1993)[8] |
| Al-Mansoura | 0.067 (1982)[3] | Damietta Estuaries | 0.044 (1982)[3] |
|  |  |  | 616.0–720.0 (1995–1997)[10] |
| Kafr El-Zayat | 0.099 (1982)[3] | Rosetta Estuaries | 0.039 (1982)[3] |
|  | 0.019–0.074 (1994)[9] |  | 566.0–624.0 (1995–1997)[10] |
| Qarun Lake | 55.23 (97–99)[14] | Damietta Branch | 90.0–321.0 (1988)[4] |
| Rayan Lakes | 0.0 (97–99)[14] | Rosetta Branch | 23.2–431.0(1988)[(4)] |
| Edku Lake | 418.0(97–98)[12] | Irrigation/drainage | 2.21 (na)[1] |
| Maryout Lake | 4.29–9.15 (na)[2] 3.03 (na)[1] | canals | 94.3–582.9 (1996)[11] |
|  |  |  | 0.91–1.38 (1997–1998)[12] |

na, year of sampling is not available.
References (see superscripted numbers in table): 1, El-Sebae and Abo-Elamayem (1978); 2, Abo-Elamayem et al. (1979); 3, El-Dib and Badawy (1985); 4, El-Gendy et al. (1991); 5, Abou-Arab et al. (1995); 6, Badawy et al. (1995); 7, Moursy and Ibrahim (1999); 8, Yamashita et al. (2000); 9, Dogheim et al. (1996b); 10, Abbassy et al. (1999); 11, El-Kabbany et al. (2000); 12, Abbassy (2000); 13, Abbassy et al. (2003); 14, Mansour and Sidky (2003).

of residues recorded was 616–720 ppb (Damietta) and 566–624 ppb (Rosetta). Samples taken in 1988 indicated lower residue levels in water collected from the two major branches of the River Nile. The range of the 1988 values for the Damietta and Rosetta branches was, respectively, 90–321 ppb and 23.2–431.0 ppb (Fig. 3).

El-Dib and Badawy (1985) also collected and analyzed water samples during 1982 from five different locations along the River Nile (see Table 8). Samples taken from Kafr El-Zayat contained the highest concentration levels of OCPs. A factory located at this city was a major producer of DDT and other chlorinated pesticides. The locations from which water samples were analyzed displayed a contamination hierarchy as follows: Kafr El-Zayat (0.099 ppb) > Al-Mansoura (0.067 ppb) > Damietta (0.044 ppb) > Rosetta (0.039 ppb) > Cairo (0.015 ppb) (Fig. 4). The contaminating compounds included BHC, lindane, endrin, DDTs, and PCBs.

In 1995, a large-scale monitoring program was conducted on OC residue levels in water samples collected at 15 locations along the River Nile (Wahaab and Badawy 2004). The authors concluded that residue levels (Table 9) were still within the permissible (safe) limits for drinking water (WHO 1994).

Among 27 pesticides targeted for analysis, only 5 (atrazine, diazinon, malathion, tribufos, and a chlorotriazine metabolite, DEA) were detected in drinking water samples collected from greater Cairo (Table 10). No significant differences in residues were observed between raw and filtered water samples. Residue levels, in all cases, were below drinking water and "harm to aquatic life" thresholds, indicating relatively small human and ecological risks (Potter et al. 2007).

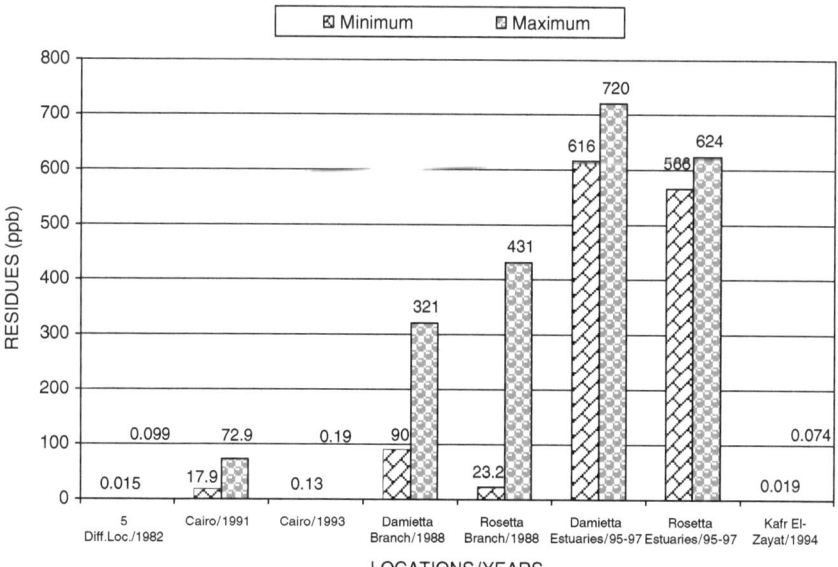

**Fig. 3** Residues (ppb) of organochlorine pesticides (OCPs) in water samples collected between 1982 and 1997 at different locations along the River Nile. (Data taken from Table 8.)

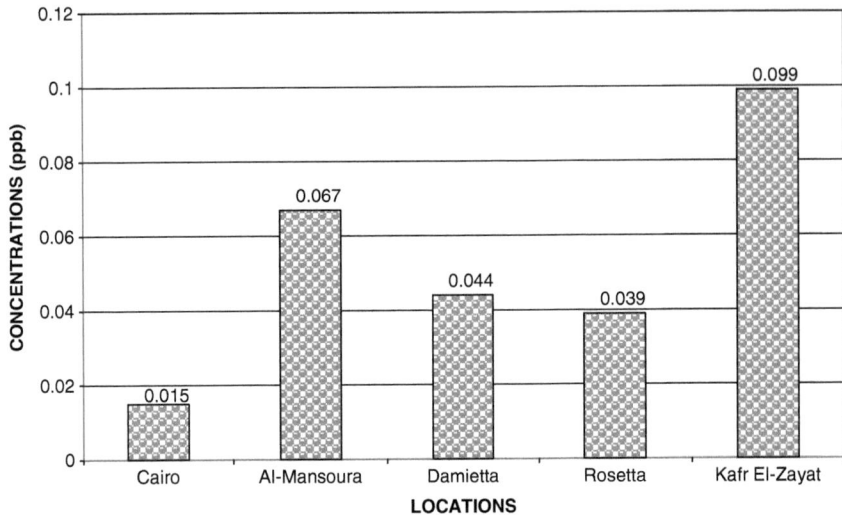

**Fig. 4** OCP water residues at different points along the River Nile in 1982. (Adapted from El-Dib and Badawy 1985.)

**Table 9** Organochlorine (OC) residues (ng/L) in water samples collected in 1995 from different locations along the River Nile[a]

| Sampling site | ΣHCHs | ΣHCBs | ΣDDTs | Cyclodienes | ΣOCs | PCBs |
|---|---|---|---|---|---|---|
| Lake Nasser | 650.46 | 81.30 | 841.47 | 20.86 | 1,594.10 | 59.86 |
| Aswan | 220.36 | 36.66 | 1,048.24 | 28.50 | 1,333.80 | 56.38 |
| Kom Ombo | 187.71 | 31.16 | 1,035.25 | 41.27 | 1,295.40 | 85.65 |
| Esna | 177.62 | 32.20 | 586.30 | 75.37 | 871.49 | 15.65 |
| Naga Hammady | 123.16 | 24.56 | 297.72 | 40.65 | 486.09 | 32.84 |
| Assuit | 143.65 | 28.37 | 100.56 | 75.28 | 247.86 | 58.46 |
| El-Menia | 163.76 | 30.35 | 82.42 | 16.77 | 293.30 | 28.58 |
| Beni Suef | 285.40 | 42.85 | 56.83 | 26.50 | 411.58 | 25.30 |
| Delta barrage | 22.00 | 10.00 | 2.65 | 29.75 | 64.40 | 8.28 |
| Kafr El-Zayat | 249.34 | 49.65 | 29.75 | 415.37 | 744.16 | 652.84 |
| Dessouk | 166.27 | 53.56 | 37.21 | 186.82 | 443.86 | 295.46 |
| Edfina | 107.26 | 77.80 | 10.13 | 228.11 | 423.30 | 71.76 |
| Rosetta | 185.87 | 16.70 | 98.51 | 32.39 | 333.42 | 140.52 |
| Al-Mansoura | 151.01 | 92.61 | 102.67 | 48.43 | 394.72 | 32.43 |
| Damietta | 26.12 | 3.90 | 90.87 | 65.57 | 186.46 | 73.66 |

[a]Levels of OC insecticides in the River Nile water are within permissible limits for drinking water (WHO 1994).
*Source*: Adapted from Wahaab and Badawy (2004).

Other studies have shown that the hazard of water contaminants can also be assessed and monitored using certain aquatic invertebrates such as the microcrustacean, *Daphnia magna* Straus, and the mosquito larva, *Culex pipiens* L. (Mansour 1998; Mansour et al. 2001a).

**Table 10** Concentration (ppb) of pesticides in samples of raw and filtered drinking water collected from Greater Cairo in December 2004

| Pesticide[a] | Median ppb (µg/L) | | Detects (%) | |
|---|---|---|---|---|
| | Raw water | Filtered water | Raw water | Filtered water |
| Atrazine | 0.005 | 0.005 | 92 | 92 |
| DEA | 0.003 | 0.001 | 75 | 50 |
| Diazinon | <0.001 | <0.001 | 42 | 33 |
| Malathion | 0.006 | 0.006 | 58 | 58 |
| Tribufos | 0.02 | 0.02 | 100 | 92 |
| Carbaryl | <DL | <DL | – | – |
| Chlorpyrifos | <DL | <DL | – | – |

DEA, chlorotriazine metabolite; DL, detection limit of 0.01 µg/L.
[a]Selected from among 27 targeted analytes representing commonly used insecticides, fungicides, herbicides, and defoliants.
*Source*: Adapted from Potter et al. (2007).

## Contamination of Sediments

Coastal sediments act as temporary or long-term sinks for many classes of anthropogenic contaminants. During the past decade, environmental regulation has reduced the loading of wastes from terrestrial sources. OCs such as polychlorinated biphenyls (PCBs) and chlorinated pesticides represent an important group of persistent organic pollutants (POPs) that have caused worldwide concern because of their toxicity and propensity to contaminate the environment. These compounds persist in sediments and may serve as non-point source reservoirs. Such sediments may release OCs over many years, even after their use ends, and the OCs may cause adverse effects in organisms and to human health through trophic transfer (Hong et al. 1995). Many POPs are believed to be possible carcinogens or mutagens and have raised concerns for human and environmental health (Tanabe et al. 1982; Allen-Gill et al. 1998; Wade et al. 1998).

Sediment samples collected from several locations in Egypt showed variable concentration levels of OC compounds (Table 11). Samples collected in 1991 from Lake Manzala and the River Nile at Cairo were found to contain residue values ranging from 661.0 to 900.0 ppm and 627.0 to 981.0 ppm, respectively (Abou-Arab et al. 1995), compared to 190.0 ppm and 100.0 ppm, as mean residue values from these sites, respectively, for 1993 (Yamashita et al. 2000).

Figure 5 clearly demonstrates that OCP residues in sediments reached a peak in 1991, preceded by very low levels (0.004–0.550 ppm) during 1979–1988, and followed by declining levels during 1993–1999. Rayan Lakes contained the lowest concentration values (0.0001 ppm; see Table 11). Rayan Lakes, created in 1973, are man-made lakes in the western desert of Egypt. These lakes were designed to receive agricultural drainage waters and are considered to be virgin ecosystems (Mansour and Sidky 2003).

**Table 11** Organochlorine residues in sediment samples and year(s) collected from different locations in Egypt

| Location | Concentration (ppm) | Location | Concentration (ppm) |
|---|---|---|---|
| Lake Manzala | 661.0–900.0 (1991)[3] | River Nile | |
| | 190.0 (1993)[5] | El-Menia | 0.0028–0.0031 (1979)[1] |
| Alexandria Harbor | 0.50–929.0 (1999)[8] | Dakahleih | 0.0053–0.0062 (1979)[1] |
| Abu-Quir Bay | 300.0 (na)[7] | Beheira | 0.0020–0.0024 (1979)[1] |
| El-Max Bay | 195.0 (?)[7] | Cairo | 627.0–981.0 (1991)[3] |
| | | | 100.0 (1993)[5] |
| | | | 0.009–0.113 (1997–1998)[10] |
| Maryout Lake | 0.371–1.13 (?)[2] | Kafr El-Zayat | 0.01–0.37 (1994)[9] |
| | 30.0 (na)[6] | | |
| Qarun Lake | 11.61 (1997–1999)[11] | Damietta Branch | 0.37 (1988)[4] |
| Rayan Lakes | 0.0001 (1997–1999)[11] | Rosetta Branch | 0.55 (1988)[4] |

na, year of sampling is not available.
References (see superscripted numbers in table): 1, Aly and Badawy (1981); 2, Abo-Elamayem et al. (1979); 3, Abou-Arab et al. (1995); 4, El-Gendy et al. (1991); 5, Yamashita et al. (2000); 6, Barakat (2004); 7, Abd-Allah and Abbas (1994); 8, Barakat et al. (2002); 9, Dogheim et al. (1996b); 10, Abbassy et al. (2003); 11, Mansour and Sidky (2003).

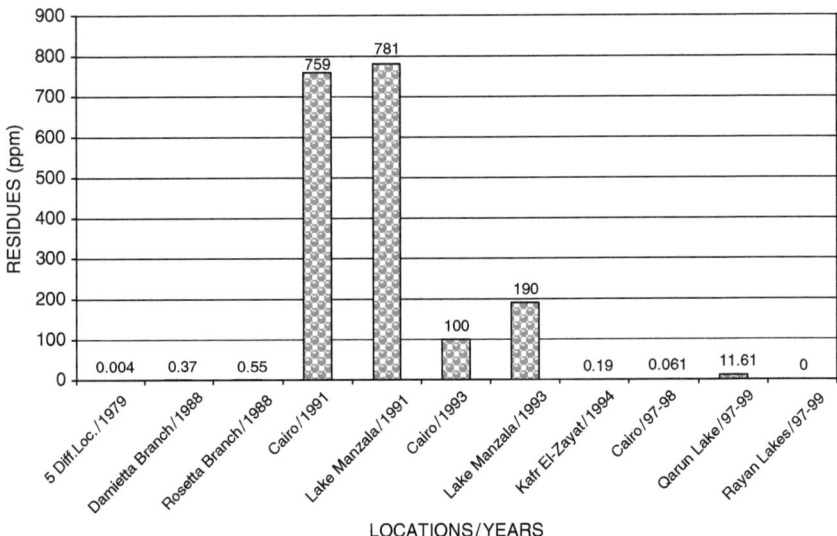

**Fig. 5** Residue levels (ppm) of OCPs in sediment samples from several Egyptian aquatic ecosystems (1979 to 1999). (Data taken from Table 11.)

## Contamination of Fish

Fish are often at the top of the aquatic food chain and may concentrate large amounts of some metals and persistent organic compounds, which accumulate differentially in fish organs and may cause serious health hazards to humans. For this reason, the problem of fish contamination by toxic substances has received great attention from several workers (e.g., Barak and Mason 1990; Sarkar and Everaarts 1998).

Extensive studies have been performed in the last 20 yr on fish samples from the River Nile and its branches in the delta region and from coastal lakes. The latter are known to receive municipal, industrial, and agricultural wastewaters. Most data are for chlorinated pesticides (e.g., cyclodiene insecticides, DDTs, HCHs, and HCB) and PCBs, whereas there is a paucity or absence of data for other persistent toxic substances (Barakat 2004).

Table 12 shows residue levels of OCPs in fish samples collected from several lakes and the River Nile. It appears that contamination levels in fish from Lake Manzala are nearly stable at approximately 45 ppb over the period 1991–1998. Generally, results indicate that contamination in the River Nile fish is higher than in the fish taken from lakes. Data illustrated in Fig. 6 reveal that OCP contamination levels in fish collected at the same time from different locations along the River Nile, in descending order, were Al-Mansoura (199.9 ppb), Assuit (176.0 ppb), Faraskur (144.2 ppb), Edfina (128.0 ppb), Cairo (123.6 ppb), and Aswan (103.5 ppb) (Aly and Badawy 1984).

**Table 12** Organochlorine residues in fish taken from Egyptian waters

| Location | Concentration (ppb) | Location | Concentration (ppb) |
|---|---|---|---|
| Lake Manzala | 32.0–60.0 (1991)[1] | S. Sinai and Suez | 9.0–66.0 (2001)[8] |
|  | 40.0 (na)[5] | Port-Said and | 10.0–210.0 (2001)[8] |
|  | 53.0 (1993)[6] | Damietta |  |
|  | 5.0–90.0 (1997–1998)[9] |  |  |
| Edku Lake | 71.0 (na)[2] | River Nile |  |
|  | 10.0 (na)[4] | Aswan | 103.5 (1981–1982)[12] |
| Maryout Lake | 90.0 (na)[2] | Assuit | 176.0 (1981–1982)[12] |
|  | 12.0 (na)[3] |  |  |
| Abu-Quir Bay | 171.0 (na)[2] | Cairo | 34.0–65.0 (1991)[1] |
|  | 55.0 (na)[4] |  | 123.6 (1981–1982)[12] |
| El-Max Bay | 80.0 (na)[3] | Al-Mansoura | 200.0 (1981–1982)[12] |
| Qarun Lake | 129.0–640.0 (1997–1999)[11] | Edfina | 128.0 (1981–1982)[12] |
| Rayan Lakes | 28.0–300.0 (1997–1999)[10] | Faraskur | 144.2 (1981–1982)[12] |
| Temsah Lake | 13.3–287.5(na)[13] | Kafr El-Zayat | 800.0 (1994)[7] |

na, year of sampling is not available.
References (see superscripted numbers in table): 1, Abou-Arab et al. (1995); 2, El-Nabawy et al. (1987); 3, Abd-Allah and Ali (1994); 4, Abd-Allah (1994) Badawy and Wahaab (1997); 6, Yamashita et al. (2000); 7, Dogheim et al. (1996b); 8, El-Nemr and Abd-Allah (2004); 9, Abbassy et al. (2003); 10, Mansour and Sidky (2003); 11, Mansour et al. (2001b); 12, Aly and Badawy (1984); 13, Ahmed et al. (2001b).

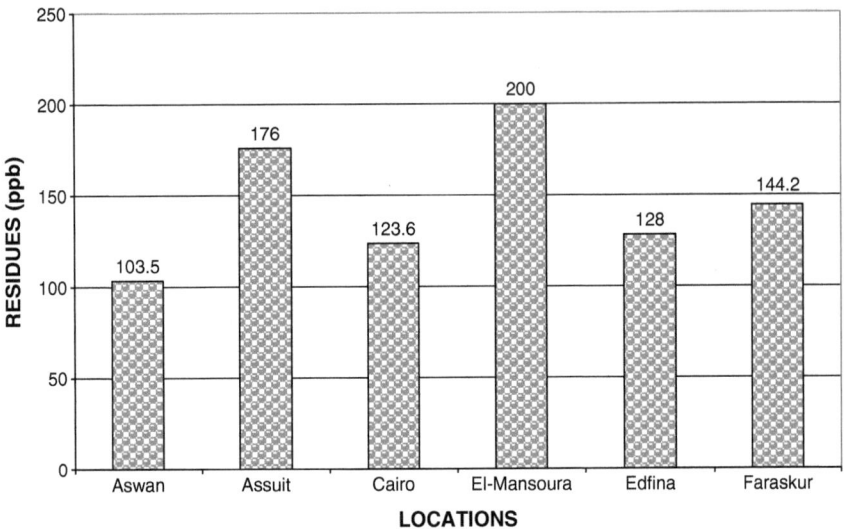

**Fig. 6** Levels of OCP residues (ppb) in fish from several locations along the River Nile (1981–1982). (Adapted from Aly and Badawy 1984.)

Table 13 includes data on OCP residues in four fish species obtained from four Governorates (South Sinai, Suez, Port-Said, and Damietta). Concentrations of OCP residues in these fish, in descending order, were Bouri > Denis > Moza > Mousa. In summary, fish from the Mediterranean Sea were more polluted than those from the Red Sea (Table 14) (El-Nemr and Abd-Allah 2004).

According to Khaled et al. (2004), the range of OC residues in bivalve mussels (*Brachiodontes* sp.) from 11 sampling locations along the Red Sea Coast were DDTs (125.0–772.0 ng/g wet wt), PCBs (6.7–66.4 ng/g wet wt), HCHs (12.6–188.4 ng/g wet wt), and cyclodienes (8.8–221.6 ng/g wet wt). With the exception of rather high results from the DDTs, these results, when compared to data from other coastal areas, indicated low to moderate PCB and insecticide contamination in mussels.

## Distribution of OCP Residues in Aquatic Ecosystems

Contamination of the aquatic environment by chlorinated hydrocarbons is of great concern because their residues reside in multiple compartments of the aquatic ecosystem. OCs share common physical and chemical properties, such as high chemical stability, interactivity, appreciable volatility at ambient temperatures (Mackay and Wolkoff 1973; Mackay and Leinonen 1975), and low water solubility (Mackay et al. 1980). However, the OC compounds have high lipid–water partition coefficients (Chiou et al. 2001). Because of these characteristics, the OCs are well known for environmental persistence, their presence in surface waters and groundwater, their ubiquitous distribution throughout the world, and accumulation in fat reserves

**Table 13** Average concentrations (ppb, wet wt) of OC pesticide residues in four fish species collected from South Sinai and Suez Governorates (Red Sea) and Port-Said and Damietta Governorates (Mediterranean Sea) in 2001

| Location/fish sample | EOM[a] | HCB | Lindane | ΣCyclodienes | ΣDDTs | Total residues[b] | C[c] (%) | D[d] (%) |
|---|---|---|---|---|---|---|---|---|
| South Sinai | | | | | | | | |
| Bouri | 77 | 20.3 | ND | 16.9 | 18.2 | 55.4 | 31 | 33 |
| Denis | 65 | 12.2 | 2.1 | 14.4 | 15.1 | 41.7 | 29 | 36 |
| Moza | 47 | 5.4 | ND | 4.4 | 7.2 | 17.0 | 26 | 42 |
| Mousa | 33 | ND | ND | 1.4 | 7.1 | 8.5 | 17 | 83 |
| Suez | | | | | | | | |
| Bouri | 54 | 33.8 | 2.4 | 8.8 | 20.1 | 66.2 | 13 | 30 |
| Denis | 45 | 20.1 | 2.1 | 4.4 | 11.6 | 38.2 | 12 | 30 |
| Moza | 39 | 10.2 | ND | 2.1 | 9.9 | 22.2 | 10 | 45 |
| Mousa | 35 | 3.5 | ND | 2.7 | 3.3 | 9.5 | 28 | 35 |
| Average | | 13.2 | 0.8 | 6.9 | 11.6 | 32.5 | 21 | 36 |
| Port-Said | | | | | | | | |
| Bouri | 62 | 15.2 | 10.1 | 14.9 | 28.4 | 68.6 | 22 | 42 |
| Denis | 55 | 8.4 | ND | 8.4 | 23.6 | 40.4 | 21 | 58 |
| Moza | 38 | 5.4 | 7.6 | 4.2 | 9.2 | 26.4 | 16 | 45 |
| Mousa | 42 | 2.5 | ND | 1.5 | 5.5 | 9.5 | 16 | 58 |
| Damietta | | | | | | | | |
| Bouri | 65 | 38.6 | 30.2 | 28.0 | 91.3 | 188.1 | 15 | 49 |
| Denis | 60 | 17.4 | 11.2 | 15.0 | 46.2 | 89.8 | 17 | 51 |
| Moza | 50 | 9.2 | 7.4 | 6.5 | 16.5 | 39.6 | 16 | 42 |
| Mousa | 47 | 5.1 | ND | 1.1 | 8.4 | 14.6 | 8 | 58 |
| Average[e] | | 12.7 | 8.3 | 10.0 | 28.6 | 59.6 | 16.4 | 49.1 |

[a] EOM, extractable organic matter.
[b] Total residues = sum of concentrations of all estimated pesticide residues.
[c] C (%), total cyclodienes/total residues.
[d] D (%), total DDTs/total residues.
[e] Total concentrations of the 10 OCs found in fish, in descending order, were Bouri > Denis > Moza > Mousa.
Source. Adapted from El-Nemr and Abd-Allah (2004).

**Table 14** Summary of total OC residues found in fish[a] taken from the Red Sea and the Mediterranean Sea[b]

| | Residues in fish (ng/g wet wt) | | | |
|---|---|---|---|---|
| Location | Bouri | Denis | Moza | Mousa |
| RED SEA | | | | |
| South Sinai | 55.4 | 41.7 | 17.0 | 8.5 |
| Suez | 66.2 | 38.2 | 22.2 | 9.5 |
| MEDITERRANEAN SEA | | | | |
| Port Said | 68.6 | 40.4 | 26.4 | 9.5 |
| Damietta | 210.5 | 101.9 | 39.6 | 18.1 |

[a] Four kinds of fish were studied: Bouri, Denis, Moza, and Mousa.
[b] Generally, fish from the Mediterranean Sea were more polluted than those from the Red Sea.
Source: Adapted from El-Nemr and Abd-Allah (2004).

of marine and terrestrial organisms. Table 15 shows that the distribution of OCP residues in water, fish, and sediment samples collected in 1991 from Lake Manzala were 0.106, 0.05, and 780.0 ppm, respectively. The distribution of residues in these three sample types, collected in 1993, were 0.0014, 0.053, and 190.0 ppm, respectively; the pattern seen in samples collected in 1997–1998 was 0.00013, 0.03, and 0.05 ppm, respectively. These results show that the OCs are preferentially distributed into these ecosystem compartments. According to Sarkar and Everaarts (1998), the highest concentrations of t-DDT and t-HCH were detected in sediment (501 and 35 ng/g, respectively), followed by biota, e.g., *Solea vulgaris* (148 and 28 ng/g, respectively), and water (100 and 0.03 ng/L, respectively). In Table 16, the relative accumulation indices were calculated using data from Table 15. In general, the accumulation pattern for pesticides (and heavy metals) in different ecosystem compartments has the following hierarchy: sediment > fish> water (Mansour et al. 2001b; Mansour and Sidky 2002; Mansour 2006).

Table 17 presents available data on occurrence of PCBs in different components of the aquatic ecosystem representing River Nile branches and estuaries, lakes, and some locations on the Mediterranean and the Red Seas. The data reveal that the highest concentration of PCBs in water exists in the Nile Estuary at Damietta

**Table 15** Organochlorine pesticide residues in major components of Egypt's aquatic ecosystems

| | | Concentrations | | | |
| --- | --- | --- | --- | --- | --- |
| Location | Year of sampling | Water (ppb) | Fish (ppm) | Sediment (ppm) | Reference |
| Manzala Lake | 1991 | 106.0 | 0.05 | 780.0 | (1) |
| Cairo | 1991 | 45.4 | 0.05 | 759.0 | (1) |
| Manzala Lake | 1993 | 1.41 | 0.053 | 190.0 | (2) |
| Kafr El-Zayat | 1994 | 0.047 | 0.80 | 0.20 | (3) |
| Manzala Lake | 1997–1998 | 0.13 | 0.03 | 0.05 | (4) |
| Qarun Lake | 1997–1999 | 55.23 | 2.25 | 11.61 | (5) |

*Sources*: Adapted from (1) Abou-Arab et al. (1995); (2) Yamashita et al. (2000); (3) Dogheim et al. (1996b); (4) Abbassy et al. (2003); (5) Mansour and Sidky (2003).

**Table 16** Relative accumulation of OC pesticides in water, fish and sediment of Egyptian aquatic ecosystems

| | | Relative accumulation ($\times$ times) | | |
| --- | --- | --- | --- | --- |
| Location | Year of sampling | Water | Fish | Sediment |
| Manzala Lake[1] | 1991 | 1.00 | 0.47 | $7.4 \times 10^3$ |
| Manzala Lake[2] | 1993 | 1.00 | 37.6 | $1.3 \times 10^5$ |
| Manzala Lake[3] | 1997–1998 | 1.00 | 23.1 | 38.5 |
| Cairo[1] | 1991 | 1.00 | 1.1 | $1.7 \times 10^4$ |
| Kafr El-zayat[4] | 1994 | 1.00 | $1.7 \times 10^4$ | $4.3 \times 10^3$ |
| Qarun Lake[5] | 1997–1999 | 1.00 | 41.0 | 211.1 |

*Sources*: Adapted from (1) Abou-Arab et al. (1995); (2) Yamashita et al. (2000); (3) Abbassy et al. (2003); (4) Dogheim et al. (1996b); (5) Mansour and Sidky (2003).

**Table 17** Concentration of polychlorinated biphenyls (PCBs) in water, fish, and sediment of Egyptian aquatic ecosystems

| Location | Water Year | Water ppb | Fish Year | Fish ppm | Sediment Year | Sediment ppm |
|---|---|---|---|---|---|---|
| **River Nile** | | | | | | |
| Five different locations[a] | 1982 | 0.007–0.050[1] | – | – | 1993 | 300.0[15] |
| Rosetta Branch | 1988 | 8.0–801.0[2] | – | – | 1988 | 780.0[2] |
| Damietta Branch | 1988 | 32.0–74.0[2] | – | – | 1988 | 190.0[2] |
| **Nile estuaries** | | | | | | |
| Rosetta | 1995–1997 | 556.0–611.0[3] | – | – | – | – |
| Damietta | 1995–1997 | 851.0–1030.0[3] | – | – | – | – |
| Lake Manzala | 1992 | 0.003–0.019[4] | na | 0.017[12] | 1993 | 140.0[15] |
| Lake Manzala | 1997–1998 | 0.073–0.20[5] | 1997–1998 | 0.033–0.126[5] | 1997–1998 na | 0.021–0.620[5] 120.0[16] |
| **Mediterranean Sea** | | | | | | |
| Drainage into Mediterranean Sea | 1997–1998 | 0.69–1.54[6] | – | – | – | – |
| El-Max Bay | na | 26.0–191.0[7] | na | 0.080[11] | na | 220.0[8] |
| Abu-Quir Bay | na | 20.0–844.0[8] | na | 0.070[10] | na | 370.0[8] |
| El-Max Coast | na | 37.0–131.0[9] | na | 0.018[14] | – | – |
| Alexandria Harbor | – | – | – | – | 1999 | 0.90–1210.0[17] |
| Edku Lake | 1997–1998 | 720.0[6] | na | 0.017[13] | – | – |
| Lake Maryout | – | – | na | 0.028[11] | – | – |
| Lake Maryout | – | – | na | 0.022[14] | – | – |
| Red Sea Coast | – | – | 2000 | 6.7–66.4[18] | – | – |
| Temsah Lake | – | – | 2002 | 0.21–6.7[19] | – | – |

na, the year of sampling is not available.
[a]Locations: Cairo, Kafr El-Zayat, Rosetta, Al-Mansoura, and Damietta.
References: [1] El-Dib and Badawy (1985); [2] El-Gendy et al. (1991); [3] Abbassy et al. (1999); [4] Badawy et al. (1995); [5] Abbassy et al. (2003); [6] Abbassy (2000); [7] Abd-Allah (1992); [8] Abd-Allah and Abbas (1994); [9] Abd-Allah (1999); [10] El-Nabawy et al. (1987); [11] Abd-Allah and Ali (1994); [12] Badawy and Wahaab (1997); [13] Abd-Allah (1994); [14] El-Nabawy et al. (1987); [15] Yamashita et al. (2000); [16] Barakat (2004); [17] Barakat et al. (2002); [18] Khaled et al. (2004); [19] Tundo et al. (2005).

(851.0–1030.0 ppb). The highest concentration in fish was seen in samples from the Red Sea Coast (6.7–66.4 ppm). Sediment retained higher PCB levels than did waters. Accordingly, sediment samples from Manzala Lake, Rosetta Branch, and Damietta Branch contained PCBs residues as high as 320, 1,902, and 359 times those in water at these locations, respectively (Table 18).

**Table 18** Relative accumulation of PCBs in water, fish, and sediment from certain Egyptian aquatic ecosystems

| Location | Year | Relative accumulation (× times)[a] | | |
|---|---|---|---|---|
| | | Water | Fish | Sediment |
| Lake Manzala | 1997–1998 | 1.00 | 800 | 320 |
| Rosetta Branch | 1988 | 1.00 | – | 1,902 |
| Dameitta Branch | 1988 | 1.00 | – | 359 |

[a]Relative accumulations were calculated using data from Table 17.

There are few data on contamination of aquatic ecosystems with nonorganochlorine pesticides (e.g., OPs and carbamates). Table 19 summarizes results of investigations published between 2000 and 2006 on these non-OCPs. Generally, concentration levels were reported to be "within safety margins" of their respective permissible limits.

**Pesticide Residues in Aquatic Plants and Birds**

There are few studies on uptake and accumulation of OCPs in aquatic plants and birds. In fact, a literature survey revealed only a single publication. Because of the importance of Lake Manzala as a big fish resource in Egypt, it has received considerable attention from researchers. The lake is a habitat for numerous aquatic plant species and many migratory and resident birds. It receives direct discharge of untreated wastewater and discharges of irrigation water from drainage canals and reclaimed lands. Table 20 shows levels of OC compounds detected in muscle of three bird species collected from the Lake Manzala region (Abbassy et al. 2003). The highest level was observed for $p,p'$-DDE in muscle of *Egretta ibis* (mean, 442.5 ng/g wet wt); the lowest level was observed for Aroclor 1260 (mean, 1.11 ng/g wet wt) in muscle of *Alceab athis*. Aroclor 1254 residues predominated in the three studied plants and reached the highest level (32.68 ng/g wet wt) in *Ceratophyllum demersum*. Nearly the same concentrations of total DDTs were found in the studied plants as existed for the aroclors.

## 4.6 Food Contamination

**Food Contamination with Nonpesticidal POPs**

Food contamination is the single and most common precursor of a variety of diseases and anomalies that have long plagued mankind. It has been proposed that about 30% of human cancers are caused by low exposure to initiating carcinogenic contaminants in the diet (Tricker and Preussmann 1990). Moreover, food consumption has been identified as the major pathway for human exposure to environmental

**Table 19** Residues of organophosphorus and carbamate pesticides in water, fish, and sediment samples from locations in Egypt

| Location/pesticide | Year of sampling | Concentration (ppb) | | | Reference |
|---|---|---|---|---|---|
| | | Water | Fish | Sediment | |
| El-Haram, Giza Irrigation canals | 1996 | | No data | | El-Kabbany et al. 2000 |
| Chlorpyrifos | | 13.4–15.8 | | 16.0–23.0 | |
| Dimethoate | | 12.4 | | 9.0–32.0 | |
| Parathion | | 10.8 | | 1.2–26.2 | |
| Diazinon | | ND | | 14.3–20.8 | |
| Endosulfan | | 3.4–290.2 | | ND | |
| Carbosulfan | | 12.0–21.0 | | 18.8–35.2 | |
| Aldicarb | | 8.9–42.4 | | 25.0–48.6 | |
| Carbaryl | | 19.8–48.3 | | 18.6–20.0 | |
| North Coast, Mediterranean Sea (drainage water) | 1997–1998 | | No data | No data | Abbassy 2000 |
| Dimethoate | | 0.09–0.10 | | | |
| Malathion | | 0.06–0.07 | | | |
| Captan | | 0.08–0.09 | | | |
| Ametryne | | 0.07–0.08 | | | |
| New Damietta (major drainage canal) | 1999–2001 | | | | Abdel-Halim et al. 2006 |
| Chlorpyrifos | | 24.5–303.8 | 16.5–31.6 | 0.9–303.8 | |
| Chlorpyrifos-Me | | 21.8 | ND | 61.3 | |
| Pirimiphos-Me | | 23.3 | 3.1 | ND | |
| Profenofos | | 41.0 | 2.1–12.6 | ND | |
| Malathion | | 71.9–466.0 | 4.9–19.3 | 2.0–5.12 | |
| Diazinon | | 24.6–70.5 | 21.1–43.0 | 0.9–279.0 | |
| Western Desert Lakes (Rayan) | 1997–1999 | | | | Mansour and Sidky 2003 |
| Malathion | | ND | ND | 2.0–3.0 | |
| Pirimiphos-Me | | ND | 82.0–132.0 | ND | |
| Diazinon | | 4.5 | ND | ND | |
| Western Desert Lakes (Qarun) | 1997–1999 | | | | Mansour et al. 2001b |
| Malathion | | 14.1 | 43.8 | ND | |
| Pirimiphos-Me | | 24.8 | 50.5 | ND | |
| Profenofos | | ND | ND | ND | |

ND, not determined.

**Table 20** Mean concentration (ng/g wet wt) of xenobiotic residues in aquatic plants and birds dominant in Manzala Lake, Egypt

| Pesticide or PCB | Plants[a] | | | Birds[b] | | |
|---|---|---|---|---|---|---|
| | 1 | 2 | 3 | I | II | III |
| HCB | 3.35 | 7.08 | 1.58 | 3.27 | 2.49 | 2.26 |
| Lindane | 0.65 | 0.14 | 0.18 | 1.66 | 1.58 | 1.16 |
| $p,p'$-DDE | 5.65 | 5.09 | 2.96 | 210.9 | 14.57 | 442.5 |
| $p,p'$-DDD | 0.45 | 0.46 | 1.10 | 4.53 | ND | 1.17 |
| $p,p'$-DDT | 0.53 | 0.55 | 2.11 | 2.48 | 1.38 | ND |
| ΣDDTs | 6.63 | 6.10 | 6.17 | 217.91 | 15.95 | 443.67 |
| Aroclor 1254 | 32.68 | 8.92 | 7.02 | 79.87 | 23.79 | 15.87 |
| Aroclor 1260 | 0.38 | 0.12 | 0.43 | 2.58 | 1.11 | 1.93 |

[a]Plants:
1, *Ceratophylum demersum*
2, *Lema* spp.
3, *Potamogeton pictinatus*

[b]Birds:
I, Moorhen (*Gallinula chloropus*)
II, Kingfisher (*Alceab athis*)
III, Cattle egret (*Egretta ibis*).

*Source*: Adapted from Abbassy et al. (2003).

contaminants, accounting for more than 90% of intake compared to inhalation or dermal routes of exposure (Fries 1995). Food monitoring studies in Egypt have been primarily limited to analyses of OC, OP, and carbamate insecticides. Research on other POPs, such as polychlorinated dibenzo-*p*-dioxins and dibenzofurans (PCDD/Fs), PCBs, and PAHs is a relatively new subject in Egypt. Therefore, only recently have articles been published on residues of dioxins and furans in some environmental matrices and in marine organisms. Below, a brief account is given of work on this group of POPs.

POPs include pollutants that are semivolatile, persist in the environment, bioaccumulate, and are toxic for humans and wildlife. The main groups, namely PCDD/Fs, PCBs, and PAHs, emerge from anthropogenic causes but may also be produced from natural processes. PCDD/Fs are formed as unwanted by-products in many industrial and combustion processes. PCBs have been used commercially since 1929 in hundreds of industrial and commercial applications, and PCB residues still exist in the environment. PAHs are released into the environment as by-products of incomplete combustion of organic matter. The PAHs can potentially inflict irreparable damage on biologic systems and many of these compounds, such as benzo($\alpha$) pyrene, are known to be probable human carcinogens. Various toxic effects produced by this group, including potential endocrine and reproductive effects and carcinogenicity, are well documented.

Samples of human food (e.g., butter, seafood, meat), animal feed (chickens, cattle, and fish), as well as water and sediment, were subjected to analyses for residues of PCDD/Fs and PCBs. Butter contained levels of these contaminants several times higher than that reported in EU countries and exceeded the EU limits, whereas the PCDD/Fs levels in seafood and feed were far below current EU limits (Tundo et al. 2004, 2005; Loutfy et al. 2006, 2007).

The total concentration of PAHs in bread samples collected from Cairo ranged from 50.1 to 450.4 (mean, 187.0 µg kg$^{-1}$), compared to 56.6 µg kg$^{-1}$ in Ismailia bread (Ahmed et al. 2001a).

The total PAHs in mullet fish (*Mugil cephalus*), crab (*Lupa pelagus*), and bivalve (*Ruditapes decussate*) from Lake Temsah, Ismailia, were 2.7–11.1 (mean, 7.7 ppb) in fish, 48.9 ppb in bivalve, and 7.1 ppb in crab (Ahmed et al. 2001b).

Loutfy et al. (2006) estimated concentrations of PCDD/Fs and DL-PCBs (dioxin-like polychlorinated biphenyls) in major foodstuffs (dairy products, fish/seafood, and meat) randomly collected from Ismailia City, Egypt. To provide a primary estimation of the Toxic Equivalent (TEQ) for the main food categories consumed in Egypt, an estimation of the dietary intake (based on EU data) for measured and nonmeasured groups (e.g., cereals, vegetables/fruits, eggs, and milk) was performed. The calculated dietary intake, using WHO intake assumptions (60 kg body weight, bwt) ranged from 3.69–4.0 pg WHO-TEQ/kg bwt/day for PCDD/Fs to 6.04–6.68 pg WHO-TEQ/kg bwt/day, if DL-PCBs were included. PCBs contributed about 40% of the total TEQ intake. Dairy products (mainly cheese) were the main contributor to intake of PCDD/Fs (89%); fish/seafood and meat shared a similar percent (5.4%). The total intake (PCDD/Fs + DL-PCBs) resulting from consumption of dairy products, fish/seafood, and meat is much higher than that mentioned in recent reports from EU countries, and higher than the maximum WHO-TDI (Tolerable Daily Intake) of 4 pg TEQ/kg bwt/day. A recalculation of the dietary exposure, including all sources of intake, yielded 4.06–6.38 pg TEQ/kg bwt/day for PCDD/Fs, and a range of 6.59–9.98 pg TEQ/kg bwt/day for the total including PCBs. The authors concluded that cereals and vegetables/fruits contribute significantly to PCDD/Fs TEQ intake in Egypt and play a more important role than fish/seafood and meat.

## Food Contamination with Traditional Pesticides

Milk and Dairy Products

According to Abou-Arab (1997), concentrations of total DDT and its metabolites were found to be 0.100 mg/kg fat in raw milk samples collected from greater Cairo, compared to 0.008 mg/kg fat in Ras cheese. Also, Abou-Arab (1999) reported the following concentrations (ng/g fat) of lindane and its metabolites: raw milk (66.0); sterilized milk (23.0); yoghurt (31.0); Damietta cheese (65.0); and Ras cheese (21.0). All these levels were below the Maximum Residue Level (MRL) for lindane (10.0 μg kg$^{-1}$ fat; FAO/WHO 1993).

Data on dairy products as reported during the last decade are summarized in Table 21 (Barakat 2004). Pesticide residues in samples of buffalo milk from Beni-Suef, Fayoum, and Cairo areas revealed that most samples were contaminated with one or more OC insecticides, e.g., DDT, HCHs, heptachlor, dieldrin, HCB, and endrin (Dogheim et al. 1990; Aman and Bluthgen 1997; Amr 1999). Total DDTs exceeded the MRLs in 7 of 18 samples (Dogheim et al. 1990). The same compounds were frequently found in milk powders as in milk (Aman and Bluthgen 1997; Fahmy 1998; Amr 1999), as well as in Damietta cheese (Aman and Bluthgen 1997; Amr 1999).

**Table 21** Residue levels (ng/g dry wt lipid) of chlorinated pesticides in Egyptian dairy products

| Product | Cyclodienes | ΣDDTs | HCHs | Reference |
|---|---|---|---|---|
| Buffalo milk | 150.0–310.0 | 610.0–4260.0 | 140.0–1950.0 | Dogheim et al. 1990 |
| Buffalo milk | <DL[a] | 7.52–20.01 | 4.4–11.0 | Dogheim et al. 1996b |
| Buffalo milk | (6.10)[b] | (78.0) | (32.0) | Aman and Bluthgen 1997 |
| Buffalo milk | (324.0) | (1530.0) | (1077.0) | Amr 1999 |
| Milk powder | (10.4) | (84.0) | (20.7) | Aman and Bluthgen 1997 |
| Milk powder | – | 3.0–24.0 | 20.4–38.1 | Fahmy 1998 |
| Milk powder | (6.89) | (19.0) | (161.0) | Amr 1999 |
| Damietta cheese | (7.0) | (114.0) | (35.0) | Aman and-Bluthgen 1997 |
| Damietta cheese | (13.0) | (47.0) | (63.4) | Amr 1999 |

[a]Below detection limit (DL).
[b]Numbers in parentheses are mean values.
*Source*: Adapted from Barakat (2004).

## Meat and Cereals

El-Kady et al. (2001) measured pesticide residues in local and imported beef (64 meat and 18 liver samples). Based on the data presented in Table 22, the authors concluded that 22.2% of local liver samples contained lindane residues above the MRLs of the Egyptian Organization for Standardization and Quality Control (EOS 1991) and Codex (1993). Only 11.1% of the imported samples contained these residues. Also, 3.1% of local meat samples contained pirimiphos-methyl (587.6 μg/kg fat), which exceeded its MRL (50.0 μg/kg fat). Similarly, one positive local meat sample contained profenofos (63.3 μg/kg fat) which is also above its MRL (20.0 μg/kg fat).

Twenty kilograms each of white corn and wheat grains were collected from local markets in Cairo and subjected to pesticide residue analyses (Salim and Zohair 2004). Results (Table 23) indicated that 97% of samples analyzed were contaminated with different types of OC and OP compounds. Malathion, pirimiphos-methyl, and chlorpyriphos-methyl were found in corn at levels of 2.7, 8.2, and 5.3 mg/kg and in wheat at 3.0, 4.1, and 6.0 mg/kg, respectively. Such high levels of OP insecticides in corn and wheat may have resulted from postharvest treatments.

## Vegetables and Fruits

Pesticide residues in certain foods have been extensively determined in Egypt. A market basket survey was conducted in 1991 to monitor OCP residues in potatoes and in citrus fruits collected from Egyptian markets (Dogheim et al. 1996a). No OC

**Table 22** Distribution of pesticides (μg/kg fat) in meat and liver samples taken from local or imported cattle

| | MEAT[a] | | | | |
| --- | --- | --- | --- | --- | --- |
| | Local (n = 32 samples) | | Imported (n = 32 samples) | | |
| Pesticide | Mean | Frequency (%) | Mean | Frequency (%) | MRL[b] |
| Lindane | 33.4 | 88.1 | 42.0 | 86.7 | 1,000 |
| ΣDDTs | 67.0 | 93.1 | 38.3 | 76.7 | 5,000 |
| Endrin | 26.9 | 14.1 | ND | – | 100 |
| Aldrin + dieldrin | 39.9 | 90.3 | 31.1 | 86.7 | 200 |
| Heptachlor epoxide | 17.8 | 25.8 | 11.0 | 30.0 | – |
| Malathion | 114.8 | 34.7 | 32.5 | 6.2 | 4,000 |
| Pirimiphos-Me | 203.8 | 5.5 | 40.0 | 3.1 | 50 |
| Profenofos | 21.1 | 2.8 | 2.6 | 3.1 | 20 |

| | LIVER[c] | | | | |
| --- | --- | --- | --- | --- | --- |
| | Local (n = 9 samples) | | Imported (n = 9 samples) | | |
| Pesticide | Mean | Frequency (%) | Mean | Frequency (%) | MRL |
| Lindane | 872.5 | 100.0 | 327.7 | 100.0 | 1,000 |
| ΣDDTs | 253.2 | 100.0 | 169.7 | 44.4 | 5,000 |
| Endrin | ND | – | 70.8 | 11.1 | 100 |
| Aldrin + dieldrin | 112.6 | 66.7 | 147.3 | 66.7 | 200 |
| Heptachlor epoxide | 59.9 | 22.2 | 100.5 | 11.1 | – |

MRL, Maximum Residue Level.
[a]*Author's comment*: 3.1% of local meat samples contained pirimiphos-methyl (587.6 μg/kg fat) exceeding the MRL (50.0 μg/kg fat). Similarly, one positive local meat sample contained profenofos (63.3 μg/kg fat) above the MRL (20.0 μg/kg fat).
[b]MRL from Egyptian Standards (Nos. 1954, 1966, 2079, 2081, 2718).
[c]*Author's comment*: 22.2% of local liver samples contained lindane above the MRLs set by the Egyptian Standards, corresponding to 11.1% of the imported samples.
*Source*: Adapted from El-Kady et al. (2001).

**Table 23** Pesticide residues in white corn and wheat grains collected from local Cairo markets in 2003

| | Mean concentration[a] (mg/kg) | |
| --- | --- | --- |
| Pesticide | White corn | Wheat |
| Malathion | 2.7 ± 0.7 | 3.0 ± 0.5 |
| Pirimiphos-Me | 8.2 ± 2.0 | 4.1 ± 1.2 |
| Chlorpyrifos-Me | 5.3 ± 1.8 | 6.0 ± 2.0 |
| Aldrin | 0.13 ± 0.03 | 0.10 ± 0.04 |
| Dieldrin | 0.05 ± 0.01 | 0.06 ± 0.02 |
| Lindane | 0.23 ± 0.10 | 0.22 ± 0.10 |
| HCB | 0.07 ± 0.02 | 0.09 ± 0.02 |
| Total DDTs | 0.267 ± 0.072 | 0.269 ± 0.077 |

[a]*Author's comment*: 97% of samples analyzed contained detectable residues of the above pesticides. The high OP insecticide levels in corn and wheat may have resulted from postharvest treatments.
*Source*: Adapted from Salim and Zohair (2004).

**Table 24** Pesticide residues in potato ($n = 54$) and citrus ($n = 53$) samples collected from local Egyptian markets during March 1991–January 1992

| Pesticide | Mean (ppm) | |
|---|---|---|
| | Potatoes[a] | Citrus[b] |
| Total HCH | 0.053 | ND |
| Cyclodienes | 0.036 | ND |
| Total DDT | 0.054 | ND |
| Chlorpyrifos | 0.028 | 0.026 |
| Dimethoate | 0.011 | 0.132 |
| Malathion | 0.212 | 0.024 |
| Pirimiphos-methyl | 0.507 | 0.036 |
| Fenitrothion | 2.220 | 0.332 |
| Parathion-methyl | 0.465 | ND |
| Profenofos | 0.508 | 0.038 |

[a] Eight potato samples exceeded the MRL for γ-HCH and two samples exceeded the DDT MRL.
[b] Citrus samples were free of OC pesticide residues.
*Source*: Adapted from Dogheim et al. (1996a).

residues were found in citrus samples. Almost all potato samples were contaminated with DDT and HCH isomers; heptachlor, endrin, and dieldrin were found less frequently (Table 24). The MRLs for γ-HCH in potatoes were exceeded in 8 of 54 samples and for DDT in 2 of 54 samples. Aging of HCH and DDT residues indicated recent illegal use of both pesticides during the potato storage period between growing seasons. Also, OCP residues were not detected in most fruits and vegetable samples collected from local markets in Cairo in 1995 (Dogheim et al. 1999). The average content in Egyptian tomatoes of HCB, lindane, dieldrin, and heptachlor epoxide was in the range of 3–9 ng/g, whereas DDT (83 ng/g) was reported at higher levels (Abou-Arab 1999).

Table 25 summarizes results of pesticide contamination levels between 1995 and 1999 in vegetables and fruits from monitoring studies at different locations in Egypt (Dogheim et al. 1999, 2001, 2002, 2004). The percentage of samples contaminated by year was 42.1%, 26.5%, 20.5%, and 23.2% for samples of 1999, 2001, 2002, and 2004, respectively.

A total of 78 vegetable samples (e.g., tomato, eggplant, cucumber, and potato) and 44 fruit samples (e.g., apple, grape, and orange) were collected from markets in Alexandria, and subjected to pesticide residue analyses (Abbassy 2001). No OCPs were detected in any of the samples analyzed. Those pesticides that were detected did not violate permitted levels in any of the samples analyzed (Table 26).

Residues of OCPs were determined in three kinds of organically farmed vegetables (green onion, beetroot, potatoes), as well as in their corresponding soils (Salim 2006), from samples collected in Sadat City, Egypt. Results (Table 27) indicate that OCP residues in the soil ranged between 9.94 and 10.82 μg/kg. Organic vegetables showed detectable residues ranging from 3.49 to 5.61 μg/kg, levels significantly below the MRLs (FAO/WHO 1993).

In comparison, Zohair et al. (2006) analyzed PAHs, PCBs, and OCPs in three carrot and four potato varieties obtained from organic farms in England, UK, as

**Table 25** Levels of pesticide contamination in vegetables and fruits collected between 1995 and 1999 from different locations in Egypt

| Commodity | Location | Year of sampling | No. of types | Total no. of samples | Contaminated samples | | | | Reference |
|---|---|---|---|---|---|---|---|---|---|
| | | | | | N 1 | % | N 2 | % | |
| Leafy vegetables | Greater Cairo, Beni Suef, Menofiya, Ismailia | 1999[a] | 11 | 444 | 103 | 23.2 | 28 | 6.3 | (1) |
| Vegetables | Greater Cairo, Beni Suef, Menofiya, Ismailia | 1977[b] | 23 | 1654 | 266 | 16.1 | 30 | 1.8 | (2) |
| Fruits | | | 16 | 664 | 208 | 31.3 | 15 | 2.3 | |
| Grand total | | | 39 | 2318 | 474 | 20.5 | 45 | 1.9 | |
| Vegetables | Greater Cairo, Beni Suef, Menofiya, Ismailia | 1996[b] | 23 | 1092 | 214 | 19.6 | 24 | 2.2 | (3) |
| Fruits | | | 15 | 487 | 205 | 42.1 | 17 | 3.5 | |
| Grand total | | | 38 | 1579 | 419 | 26.5 | 11 | 2.6 | |
| Vegetables | Greater Cairo, Beni Suef, Menofiya, Ismailia | 1995[c] | 12 | 238 | 80 | 33.6 | 3 | 1.3 | (4) |
| Fruits | | | 8 | 159 | 87 | 54.7 | 4 | 2.5 | |
| Grand total | | | 20 | 397 | 167 | 42.1 | 7 | 1.8 | |

N 1, number of all contaminated samples; N 2, number of violative samples within N 1.
[a] No OC pesticides were included in this group. The targeted pesticides (26) were mostly OPs, carbamates and triazines. Bendiocarb, chlorpyrifos, acephate metabolites, and profenofos were the pesticide residues most frequently found.
[b] OC compounds (e.g., bromopropylate and chlorothalonil) were detected in most of the samples analyzed. DDT complex, HCH isomers, and cyclodienes were not detectable in any of the samples analyzed.
[c] In addition to bromopropylate and chlorothalonil, DDT isomers were detected in carrots and eggplants only. No other OCs were detected.
References: (1) Dogheim et al. (2004); (2) Dogheim et al. (2002); (3) Dogheim et al. (2001); (4) Dogheim et al. (1999).

well as on their corresponding soils. Results (Table 28) reveal that concentrations of the three classes of studied pollutants in the soils ranged from 590 to 2301 µg/kg, 3.56 to 9.61 µg/kg, and 52.2 to 478.0 µg/kg, respectively. Residue uptake for carrot varieties was 8.42–40.1 µg/kg for ΣPAHs, 0.83–2.68 µg/kg for ΣPCBs, and 8.09–133.0 µg/kg for ΣOCPs. Residue uptake for potato varieties was 8.50–33.68 µg/kg for ΣPAHs, 1.04–2.68 µg/kg for ΣPCBs, and 10.36–88.36 µg/kg for ΣOCPs. Residue uptake from soils depended on plant variety. Peeling either carrots or potatoes removed 52%–100% of the contaminant residues, which varied with crop variety and the properties of the contaminants.

**Table 26** Pesticide residues (ppb) in vegetables[a] and fruits[a] collected from Alexandria City during 1997–1998

Mean pesticide residues (in ppb)[b]

| Commodity | Cypermethrin | Dimethoate | Profenofos | Dithiocarbamates (as Zineb) | Triazophos | Fenitrothion | Malathion |
|---|---|---|---|---|---|---|---|
| I. Vegetables | | | | | | | |
| Tomato | 8.3 (20.0%) | 12.6 (13.3%) | 20.8 (33.3%) | 75.0 (73.3%) | ND | ND | ND |
| Eggplant | ND | ND | 23.4 (50.0%) | 160.0 (80.0%) | 15.0 (60.0%) | ND | ND |
| Cucumber | ND | 18.0 (20.0%) | 16.8 (30.0%) | 434 (80.0%) | ND | ND | ND |
| Potato | ND | ND | ND | ND | ND | 28.0 (50.0%) | 21.2 (39.0%) |
| II. Fruits | | | | | | | |
| Apple | ND | 7.8 (25.0%) | 9.8 (33.3%) | 28.1 (50.0%) | ND | ND | 7.0 (33.3%) |
| Grape | 15.7 (30.0%) | 18.2 (40.0%) | 13.4 (70.0%) | 37.5 (50.0%) | ND | ND | 8.6 (50.0%) |
| Orange | ND | 4.6 (16.7%) | ND | 29.8 (33.3%) | ND | ND | 3.5 (16.7%) |

*Author's note*: No chlorinated hydrocarbon pesticides were detected in any samples analyzed. No violative pesticide residues were detected.
[a]Number of samples: vegetables, 78; fruits, 44.
[b]Values in parentheses indicate % frequency of the pesticide's detection in its respective commodity.
*Source*: Adapted from Abbassy (2001).

**Table 27** Residue levels[a] (μg/kg) of OC insecticides in organically farmed vegetables[b] and soil from an organic farm at El-Sadat City, Egypt

| Pesticide | Beetroot | Soil | Green onion | Soil | Potatoes | Soil |
|---|---|---|---|---|---|---|
| Total DDT | 2.02 | 6.90 | 2.82 | 6.77 | 1.78 | 6.18 |
| Lindane | 0.61 | 1.35 | 0.84 | 1.27 | 0.60 | 1.27 |
| Aldrin | ND | ND | ND | ND | ND | ND |
| Dieldrin | 0.35 | 0.77 | 0.70 | 0.58 | 0.30 | 0.64 |
| Heptachlor | ND | 0.26 | 0.59 | 0.31 | 0.24 | 0.28 |
| Heptachlor epoxide | 0.24 | ND | ND | ND | 0.35 | ND |
| HCB | ND | ND | ND | ND | ND | ND |
| Endrin | ND | 0.71 | ND | 0.85 | ND | 0.93 |
| Endosulfan | 0.27 | 0.83 | 0.66 | 1.03 | 0.23 | 0.64 |
| Grand total | 3.49 | 10.82 | 5.61 | 10.81 | 3.50 | 9.94 |

[a]Residues in vegetables are calculated on a whole-fruit basis.
[b]Organic vegetables under investigation showed residues at levels significantly lower than the MRLs (FAO/WHO 1993).
*Source*: Adapted from Salim (2006).

**Table 28** Polyaromatic hydrocarbon (PAH), PCB, and organochlorine pesticide (OCP) residues in UK organically farmed soils and different varieties of carrots and potatoes

| Compounds | Mean residue concentrations (μg/kg) | | |
|---|---|---|---|
| | Soil | Potato varieties[a] | Carrot varieties[b] |
| PAHs | 590.0–2301.0 | 8.50–33.68 | 8.42–40.10 |
| PCBs | 3.56–9.61 | 1.04–2.68 | 0.83–2.68 |
| OCPs | 52.20–478.0 | 10.36–88.36 | 8.09–133.0 |

*Author's comment*: Residue uptake from soils depended on plant variety: Desiree potato and Nairobi carrot varieties were more susceptible to PAH contamination. Similarly, uptake of PCBs and OCPs depended on potato variety. Peeling carrots and potatoes removed 52%–100% of residues, depending on crop variety and the properties of the contaminants.
[a]Potato varieties: Cara, Valor, Kestrel, Desiree.
[b]Carrot varieties: Major, Nairobi, Autumn Kings.
*Source*: Adapted from Zohair et al. (2006).

## Pesticide Residues in Medicinal and Aromatic Plants

Pesticide residues were also determined in medicinal plants (peppermint, chamomile, anise, caraway, and tilio) that are frequently used by infants and adults (Abou-Arab et al. 1999). Contamination levels in the nonpacked samples were higher than those in the packed samples (prepared in small bags for soaking in boiled water upon use) (Table 29). Concentration levels of OP and OC residues were generally within the limits set by the EOS.

Other monitoring studies were conducted by Abou-Arab and Abou-Donia (2001) and Dogheim et al. (2004). The results are summarized in Table 30. Dogheim et al. (2004) gave the number of contaminated samples and their proportion as 286 (73.1%), and similar numbers for violated samples, 173 (44.2%). No OC pesticides were included in their analyses. Malathion, profenofos, dimethoate, and pirimicarb were the most frequently found pesticide residues.

Table 29 Comparison of pesticide residues in packed and nonpacked samples of some medicinal plants

| | Total residues (mg/kg) | | | | | | | | | | |
|---|---|---|---|---|---|---|---|---|---|---|---|
| | Peppermint | | Chamomile | | Anise | | Caraway | | Tilio | |
| Pesticide class | I[a] | II[b] | I | II | I | II | I | II | I | II |
| OPs[c] | 0.612 | 1.172 | 0.502 | 1.784 | 0.121 | 0.741 | 0.308 | 0.877 | 0.023 | 1.233 |
| t-DDT | 0.111 | 0.384 | 0.029 | 0.039 | 0.014 | 0.575 | 0.055 | 0.352 | 0.071 | 0.241 |
| OCs and cyclodienes | 0.100 | 0.870 | 0.951 | 1.354 | 0.0 | 1.900 | 0.151 | 2.153 | 0.0 | 0.579 |
| Grand total | 0.823 | 2.426 | 1.482 | 3.177 | 0.135 | 3.216 | 0.514 | 3.382 | 0.094 | 2.053 |

*Author's comment*: Malathion, dimethoate, and profenofos predominated in most of the analyzed samples. Concentration levels of OP and OC residues were generally within the limits of the Egyptian Organization for Standardization and Quality Control (EOS). Contamination levels in the nonpacked samples were higher than those in the packed samples.
[a]I, packed samples (prepared in small bags for ease of soaking in boiled water).
[b]II, nonpacked samples.
[c]OPs denote organophosphorus pesticides.
*Source*: Adapted from Abou-Arab et al. (1999).

**Table 30** Pesticides frequently found in medicinal and aromatic plants obtained from local Egyptian markets

| Commodities | Locations | Year of sampling | No. of samples analyzed | Pesticides most frequently found | Reference |
|---|---|---|---|---|---|
| Anise, chamomile, dry coriander, fennel | Greater Cairo, Beni Suef, Menofiya, Ismailia | 1999 | 391 | Malathion, profenofos, dimethoate, pirimicarb | Dogheim et al. 2004 |
| Spices and medicinal plants | Greater Cairo | 1999 | 303 | Malathion, dimethoate, lindane, aldrin, dieldrin, chlordane, endrin | Abou-Arab and Abou-Donia 2001 |

The data provided by Abou-Arab and Abou-Donia (2001) indicate that the number of pesticides targeted was 22, including OPs and OCs. Malathion was the most common analyte. Malathion levels in Jew's mallow (molukhia; 0.52 ppm), dill (1.72 ppm), celery (0.46 ppm), tea (0.61 ppm), caraway (0.63 ppm), chamomile (2.19 ppm), and saffron (0.67 ppm) exceeded the MRLs. Concentrations of dimethoate in caraway (1.76 ppm) and chamomile (1.78 ppm) were also above the MRLs. Residues of lindane (0.80 ppm), aldrin (0.17 ppm), dieldrin (0.17 ppm), t-DDT (1.3 ppm), chlordane (0.59 ppm), and endrin (0.14 ppm) in chamomile exceeded the MRLs. Residues of aldrin (0.09 ppm) and dieldrin (0.11 ppm) in karkade, and chlordane in peppermint, were higher than the MRLs (see Table 30).

## 4.7 Pesticide Residues in Human Blood and Urine

Ahmed et al. (2002) determined residues of DDE and PCBs in blood serum of fasting women at Port-Said. Subjects included 43 women diagnosed with invasive adenocarcinoma of the breast, 21 females suffering benign breast disease, and 11 normal females (control). Mean residues of DDE detected in the three examined groups were 41.0, 48.0, and 31.0 ng/g, respectively. PCB residues (based on a total of 29 congener peaks) were 54.9, 59.2, and 61.9 ng/g, for the three groups, respectively (Table 31). The authors added that residues of DDE in these populations of females were about 15 times higher than levels detected in similar female populations in Canada and The Netherlands.

A study was conducted to assess OC serum levels among premenopausal women and the corresponding risk of premenopausal breast cancer for women with high OC serum levels (Soliman et al. 2003). The authors found low levels of DDE, total DDT, and β-HCH in most subjects (69 breast cancer patients and 53 controls). The mean DDE level was 12.7 ppb for patients and 16.6 ppb for controls; the β-HCH level was 2.1 ppb for either patients or controls (Table 32). Women with no lactation history had much higher OC levels than women who had breast-fed ($P = 0.002$ for DDE). The study suggested that OC serum levels in Egyptian women are quite

**Table 31** Levels of DDE and PCBs in blood serum of fasted women from Port-Said, Egypt. 1999–2000

| Subjects: number, description | DDE[a] (ng/g) | PCBs (ng/g)[b] |
|---|---|---|
| 43 suffering from invasive adenocarcinoma of the breast | 41.0 ± 5.2 | 54.9 ± 7.3 |
| 21 suffering from benign breast disease | 48.0 ± 6.2 | 59.2 ± 5.6 |
| 11 normal females (control) | 31.0 ± 2.5 | 61.9 ± 8.3 |

[a]Residue levels of DDE were about 15 times higher than similar residues detected in Canada and The Netherlands.
[b]Based on 29 selected peaks comprising PCB congeners.
*Source*: Adapted from Ahmed et al. (2002).

**Table 32** Average DDE and β-HCH levels in blood serum from hospitalized premenopausal female breast cancer patients

| Subjects | DDE[a] (ppb) | β-HCH[a] (ppb) |
|---|---|---|
| 69 patients | 12.7 | 2.1 |
| 53 controls | 16.6 | 2.1 |

[a]*Author's comment:* OC serum levels were quite low, but indicate an effect of breast-feeding in eliminating OCs, which implies exposure to breast-fed children. The levels found were considered not to be a risk factor for breast cancer in the studied population.
*Source*: Adapted from Soliman et al. (2003).

low, but indicates a possible role for breast-feeding in eliminating body burdens of OCs, implying that breast-fed children may have had high exposures. OC serum levels were not considered to be a risk factor for breast cancer in the studied population.

Ninety-two Egyptian school children (46 girls and 46 boys), living in urban Giza, were studied for chronic exposure to OCP residues (Sherif et al. 2005). Varying levels of OCP residues were detected in urine of 36 of the 92 studied children, comprising 39% of the total. The most frequently detected residues were p,p'-DDE (36.1%) and p,p'-DDT (22.2%). α-HCH and heptachlor epoxide were not detected in the urine samples (Table 33).

## 4.8 Risks of Dietary Pesticide Residues

The process of dietary pesticide risk assessment involves three major components: estimation of pesticide residue levels, estimation of food consumption patterns, and characterization of risk based on a comparison of exposure estimates with toxicological criteria. Residues of pesticides in food have been extensively studied in Egypt, mostly in individual studies rather than in national monitoring programs. The diversity of methods poses some difficulties in comparing data for variables such as dietary intake, differences in foods sampled, different methods of determining food consumption, and different ways of measuring and reporting data.

**Table 33** Organochlorine pesticide residues in urine collected from 92 urban Giza school children (46 girls, 46 boys) in 2003[a]

| Detected pesticide/metabolite | Concentration (ppb) |
|---|---|
| Aldrin | 3.7 |
| Endrin | 4.3 |
| Dieldrin | 6.6 |
| Heptachlor epoxide | ND |
| α-HCH | ND |
| β-HCH | 4.5 |
| γ-HCH | 4.2 |
| p,p′-DDT | 41.9 |
| p,p′-DDE | 11.3 |
| p,p′-DDD | 4.2 |
| Total OCPs | 80.6 |

ND, not detected.
[a]Of the studied cases, 39% showed detectable levels of OCPs in their urine (20 girls and 16 boys).
*Source*: Adapted from Sherif et al. (2005).

Notwithstanding, an attempt has been made below to present an overview of such published residue data.

## Pesticide Intake by Breast-Fed Infants

During the first 3 mon of life, an infant consumes an average of 120 g/d of human milk per kg body wt. The volume consumed per unit weight decreases with age. By multiplying the concentration (μg/kg or μg/L) of a given contaminant in whole milk by 0.12, the approximate daily intake of the contaminant (μg/kg bwt) can be estimated, and this value, in turn, may be compared with international guidelines (Dogheim et al. 1991).

Table 34 summarizes results of estimated dietary intakes (EDIs) of OC pesticide residues by breast-fed infants versus acceptable daily intakes (ADIs) between 1987 and 1994 (Dogheim et al. 1991, 1996b; El-Hamid 1993). Only EDIs for aldrin, dieldrin, heptachlor, and heptachlor epoxide exceeded their ADI levels in 1990 samples from Menofiya (El-Hamid 1993). Similarly, aldrin and dieldrin EDIs from Cairo in 1987 samples exceeded their ADI levels (Dogheim et al. 1991).

## Dietary Intake of Pesticides from Fish

Tables 35 and 36 present results of risk assessments from OC analyses of fish sampled in 1995 from the River Nile and from Lake Manzala. Results are presented as percent of pesticide residue intake compared with FAO/WHO–ADI values (Wahaab and Badawy 2004). The authors concluded that levels of OCs in the analyzed fish were within safety margins for human consumption.

**Table 34** Estimated dietary intakes (EDIs) of OCP residues by breast-fed infants versus acceptable daily intake (ADI) values from samples collected between 1987 and 1994

| Pesticide | Year of sampling | Location | EDI[a] (μg/kg bw/d) | ADI (μg/kg bw/d) | EDI (% of ADI) | Reference |
|---|---|---|---|---|---|---|
| γ-HCH | 1987 | Cairo | 0.09 | 10.0 | 0.9 | (1) |
| (α+β)-HCH | | | 1.94 | –[b] | – | |
| DDT complex | | | 8.20 | 20.0 | 41.0 | |
| Aldrin, dieldrin | | | 1.22 | 0.10 | 1220 | |
| Heptachlor and its epoxide | | | 0.11 | 0.50 | 22.0 | |
| γ-HCH | 1990 | Menofiya | 0.76 | 10.0 | 7.6 | (2) |
| (α+β)-HCH | | | 14.03 | – | – | |
| DDT complex | | | 7.99 | 20.0 | 40.0 | |
| Aldrin, dieldrin | | | Ex. ADI[c] | 0.10 | | |
| Heptachlor and its epoxide | | | Ex. ADI | 0.50 | | |
| γ-HCH | 1994 | Cairo | <0.12 | 10.010.0 | <1.2 | (3) |
| | | Kafr El-Zayat | 0.31 | 3.10 | | |
| (α+β)-HCH | | Cairo | 23.32 | – | – | |
| | | Kafr El-Zayat | 14.07 | – | – | |
| DDT complex | | Cairo | 11.56 | 20.0 | 57.8 | |
| | | Kafr El-Zayat | 17.19 | 20.0 | 86.0 | |
| Aldrin, dieldrin | | Cairo | ND[d] | | ND | |
| Heptachlor and its epoxide | | Kafr El-Zayat | ND | | ND | |

[a]EDI calculated by using mean values.
[b]Data are unavailable.
[c]Ex, exceeded.
[d]ND, not detected.
*Sources*: Adapted from: (1) Dogheim et al. (1991); (2) El-Hamid (1993); (3) Dogheim et al. (1996b).

**Table 35**  Assessment of human risk from consumption of OC insecticide residues in *Tilapia* sp. fish collected in 1995 from the River Nile

| Compound | ADI[a] (μg/kg bwt) | Upper Egypt (South) | Upper Egypt (North) | Greater Cairo | Damietta Branch | Rosetta Branch |
|---|---|---|---|---|---|---|
| BHC | 42 | 1.9%[b] | 2.9% | 5.3% | 1.0% | 2.8% |
| Lindane | 65 | 3.4% | 4.5% | 5.5% | 5.9% | 5.1% |
| DDTs | 350 | 74.5% | 54.5% | 34.6% | 55.2% | 59.2% |
| Heptachlor | 350 | 12.7% | 10.8% | 8.6% | 9.8% | 8.5% |

*Author's comment*: Levels of OCs in the analyzed fish are within safety margins for safe human consumption.
[a]The ADI was calculated on the basis of a 70-kg (bwt, body weight) person consuming 150 g fish (fresh wt) per day.
[b]Results are presented as percent of FAO/WHO-ADI for each corresponding pesticide residue intake value.
*Source*: Adapted from Wahaab and Badawy (2004).

**Table 36**  Assessment of human risk from consumption of OC insecticide residues in fish collected in 1995 from Lake Manzala

| Compound | ADI[a] (μg/kg bwt) | *Clarius* sp. | *Anguilla* sp. | *Tilapia* sp. | *Mugil* sp. |
|---|---|---|---|---|---|
| BHC | 24 | 0.14%[b] | 0.30% | 0.10% | 0.26% |
| Lindane | 65 | 0.18% | 0.50% | 0.29% | 0.36% |
| DDTs | 350 | 3.10% | 0.40% | 0.80% | 2.10% |
| PCBs[c] | 200 | 0.38% | 0.84% | 0.49% | 0.56% |

*Author's comment*: Levels of OCs in the analyzed fish are within margins for safe human consumption.
[a]The ADI was calculated on the basis of a 70-kg (bwt) person consuming 150 g fish (fresh wt) per day.
[b]Results are presented as percent of FAO/WHO-ADI for each corresponding pesticide residue intake value.
[c]PCBs are calculated as Arochlor 1254.
*Source*: Adapted from Wahaab and Badawy (2004).

## Dietary Intake of Pesticides by Vegetables and Fruits

A large pesticide residue monitoring program was conducted in 1997 on vegetables and fruits. All samples (numbering 2,318) were subjected to analysis for 54 targeted pesticide residues (Dogheim et al. 2002).

Table 37 shows the pesticides most frequently detected (those found in more than two different commodities) in 10 vegetables (e.g., cucumber, eggplant, green beans, green peas, lettuce, molokhia, pepper, squash, tomato, cantaloupe) and 10 fruits (e.g., apple, apricot, dates, grape, guava, orange, peach, pears, pomegranate, strawberry). Accordingly, dimethoate was found in 9 of 10 of either vegetables or fruits (i.e., 90%). Dicofol was found in 80% of both kinds of commodities. Compounds that appeared in only 3 of 10 fruits or vegetables were considered to be among the least frequently detected pesticides (e.g., procymidone).

The general pattern of pesticide residue contamination in vegetables and fruits is shown in Table 38. About 35% of sampled peppers were contaminated with 18 different pesticides; the total mean residue value was 10.06 ppm. Among detected

**Table 37** Most frequently occurring pesticides in each of 10 selected vegetables and fruits

| Vegetables | Most frequently detected[a] | Fruits | Most frequently detected[a] |
|---|---|---|---|
| Cucumber | Dimethoate (9)[b]; dicofol (8); | Apple | Dimethoate (9)[b]; |
| Eggplant | cypermethrin (6); | Apricot | cypermethrin (8); |
| Green beans | bromopropylate (5); | Dates | dicofol (8); |
| Green peas | pirimiphos-Me (5); | Grape | malathion (7); |
| Lettuce | profenofos (5); chlorpyrifos | Guava, orange | bromopropylate (6); profenofos (5); |
| Molokhia | (4); iprodione (4); | Peach | procymidone (3); |
| Pepper | malathion (4); | Pears, | tetradifon (3); |
| Squash | chlorothalonil (4); | pomegranate | dithiocarbamates (3) |
| Tomato | dithiocarbamates (4); | Strawberry | |
| Cantaloupe | procymidone (3) | | |

[a] Pesticides found in more than two commodities.
[b] The number in parentheses is the numeric incidence of detection in the 10 vegetables or fruits.
*Source*: Adapted from Dogheim et al. (2002).

pesticides, only dimethoate violated residue limits. Pears and strawberries had the highest contamination incidence among fruits (70.6%). Eleven pesticides were detected in strawberries. These results lend credence to increasing public concern about potential toxicity from exposure to residues of many pesticides in food (Mumtaz 1995).

## Assessment of Dietary Risks from Pesticides in Egypt

To complete the review of dietary risk, an attempt was made to calculate or model probable exposure to pesticides in Egypt from residues in a collection of commonly consumed vegetables and fruits.

Inputs used for consumption of vegetables and fruits were the following:

- A diet consisting of seven kinds of vegetables (e.g., tomato, cucumber, eggplant, lettuce, peas, pepper, and squash) and three kinds of fruits (e.g., citrus, cantaloupe, and grapes) was selected.
- Food intake values given by the FAO Food Balance Sheets for the Middle East (WHO 2003), which are 233.0 g/person/d for vegetables and 204.4 g/person/d for fruits.
- Dietary consumption (g/person/d) was assumed to be tomato (81.5), cucumber (4.8), eggplant (6.3), lettuce (2.3), green peas (5.5), peppers (3.4), and squash (10.5), which adds to a total of 114.3 g/person/d for vegetables; for fruits, the total value was 112.2 g/person/d, for citrus (47.1), cantaloupe (49.3), and grapes (15.8).
- The so-called nonmeasured group (i.e., other food items not included among the selected 10 items) was estimated for vegetables intake as 233.0 – 114.3 = 118.7 g/person/d, and for fruit intake as 204.4 – 112.2 = 92.2 g/person/day.

**Table 38** General pattern of pesticide contamination for selected vegetables and fruits in a large monitoring program, 1997

| Commodity | Percent of contaminated samples[a] | Number of detected pesticides[b] | Total concentration levels (ppm) | | | Violative pesticides[c] |
|---|---|---|---|---|---|---|
| | | | Min. | Max. | Mean | |
| I. VEGETABLES | | | | | | |
| Cucumber | 30.4 | 12 | 1.14 | 5.87 | 2.49 | Cypermethrin; dicofol |
| Cantaloupe | 26.8 | 8 | 0.46 | 1.56 | 0.87 | – |
| Eggplant | 1.1 | 4 | 0.44 | 0.44 | 0.44 | – |
| Beans | 21.4 | 10 | 1.83 | 8.41 | 3.83 | Permethrin |
| Peas | 15.6 | 10 | 4.37 | 6.72 | 5.34 | Dimethoate; omethoate; permethrin |
| Lettuce | 7.1 | 2 | 2.90 | 2.90 | 2.90 | Dimethoate |
| Peppers | 34.8 | 18 | 5.85 | 20.33 | 10.06 | Diazinon |
| Squash | 31.3 | 14 | 1.62 | 6.36 | 2.91 | Carbosulfan |
| Molokhia | 8.0 | 2 | 0.94 | 1.31 | 1.13 | – |
| Tomato | 31.3 | 14 | 1.62 | 6.36 | 2.91 | Carbosulfan |
| II. FRUITS | | | | | | |
| Apple | 30.1 | 10 | 1.12 | 2.48 | 1.69 | – |
| Apricot | 14.3 | 2 | 0.66 | 0.66 | 0.66 | – |
| Dates | 8.9 | 3 | 1.82 | 1.87 | 1.84 | – |
| Grape | 39.3 | 15 | 9.49 | 18.08 | 12.92 | Cypermethrin; dimethoate |
| Guava | 24.5 | 7 | 0.58 | 2.26 | 1.04 | Bromopropylate; Cypermethrin |
| Orange | 44.0 | 6 | 0.70 | 3.30 | 1.32 | – |
| Peach | 28.6 | 7 | 4.47 | 7.70 | 5.26 | – |
| Pears | 70.6 | 5 | 1.30 | 2.19 | 1.56 | – |
| Pomegranate | 66.7 | 4 | 0.39 | 0.83 | 0.59 | Cypermethrin |
| Strawberry | 70.6 | 11 | 0.97 | 9.84 | 3.92 | – |

[a]Total number of samples analyzed: 2,318.
[b]Total number of analyzed pesticides: 54.
[c]Violative pesticides means that average residue values exceeded MRLs for that pesticide in at least one commodity.
*Source*: Adapted from Dogheim et al. (2002).

Residue data inputs were as follows:

- Maximum published residue values were used for each food item.
- Residues from the "measured" and "nonmeasured" food groups were added together to achieve a "sum of residues" for entry into the formula presented below. The "nonmeasured" group amounted to 118.7g for vegetables and 92.2g for fruits; herein, half the detection limit for each identified pesticide was multiplied by their respective values and added to the original result.

**Table 39** Calculated health risks for systemic effects associated with dietary exposure to pesticide residues in selected vegetables and fruits

| Pesticide | WHO-ADI (mg/kg/d) | Estimated dose (mg/kg/d) | Hazard index[a] | Health risk[b] |
|---|---|---|---|---|
| Bromopropylate | 0.03 | 0.0018 | 0.06 | No |
| Chlorprifos[c] | 0.01 | 0.0011 | 0.11 | (No) |
| Chlorothalonil | 0.03 | 0.0009 | 0.03 | No |
| Cypermethrin | 0.05 | 0.0025 | 0.05 | No |
| Dicofol | 0.002 | 0.0033 | 1.65 | Yes |
| Dimethoate | 0.002 | 0.0025 | 1.25 | Yes |
| Iprodione | 0.06 | 0.0029 | 0.05 | No |
| Procymidone | 0.10 | 0.0011 | 0.01 | No |
| Pirimiphos-Me[c] | 0.03 | 0.00096 | 0.03 | (No) |
| Malathion | 0.30 | 0.0018 | 0.006 | No |
| Profenofos | 0.01 | 0.0023 | 0.23 | No |
| Dithiocarbamates | – | 0.0028 | – | ? |

[a]The hazard index is calculated by dividing the estimated "dose" by the WHO-ADI.
[b]"Yes" indicates a positive health risk; "No" indicates no health risk from consuming diets contaminated with the respective pesticide residues.
[c]If the estimated (calculated) dose for chlorpyrifos or pirimiphos-methyl were compared with their RfDs set by the US EPA (0.0003 and 0.00008 mg/kg/d, respectively), their "Health Risk" would change to "Yes."

The "dose" (mg/kg bwt/d) of pesticide dietary intake was calculated from the following formula:

$$\text{Dose} = \frac{\text{Sum of Residues}\,(\text{mg}/\text{kg}) \times \text{Food Item}\,(\text{kg}/\text{day})}{\text{Body Weight}\,(60\,\text{kg})}$$

Finally, the "dose" calculated for each pesticide was compared with WHO-ADIs to render a "Health Risk" index (Table 39). It is concluded, from this exercise, that the "Hazard Index" for dicofol and dimethoate exceeds 1.0 (1.65 and 1.25, respectively), indicating a probable risk to health from consuming these in the calculated diet. Moreover, if the estimated dose for chlorpyrifos or pirimiphos-methyl is compared with their EPA's RfD values (0.0003 and 0.00008 mg/kg/d, respectively), these compounds would also be regarded as risks to health.

# 5 Factors Contributing to Pesticide Hazards: Humans and the Environment

## 5.1 General

In a previous publication, Mansour (2004) reported three major factors that contribute to health risks of pesticides in developing countries:

1. Inadequate governmental controls: e.g., lack of an effective pesticide registration scheme; importation of highly toxic pesticides or ones banned elsewhere;

absence of a national plan for monitoring residues in food; lack of adequate occupational safeguards and poisoning surveillance systems; and tolerating false advertising of pesticides in news media.
2. Laxity among pesticide users: e.g., misuse, or improper handling, of pesticides; ignoring re-entry and preharvest intervals; involvement of children and women in pesticide farm work; illiteracy, in the context of Good Agriculture Practices (GAP); and excessive use of household pesticides by the public.
3. Miscellaneous factors: e.g., malnutrition; infectious and parasitic diseases; multiple exposures to toxic mixtures; and poverty.

## 5.2 Specific

As mentioned, safety measures are generally poorly applied and workers lack proper knowledge or training in safe handling of pesticides in Egypt (Amr and Halim 1997). Unfortunately, workers are often not equipped with protective clothing or masks when they use toxic compounds. Moreover, the many thousands of children who collect egg masses of the cotton leafworm daily for about 40 d each season are exposed to insecticide residues on cotton leaves.

In an attempt to investigate the attitude and behavior of Egyptian farmers regarding pesticide use in one of the largest agricultural areas in Egypt (Damanhour, El-Beheira Governorate in the middle of the Delta), a questionnaire was designed and sent to growers. The main purpose was to get answers concerning levels of education, the farmers' knowledge of pesticides and the sources they used to obtain information about use and risk avoidance, as well as ways they dispose of empty pesticide containers.

The questionnaire was administered face-to-face by trained interviewers to 203 farmers selected randomly from villages adjacent to Damanhour center. The age of the participants ranged from 15 to 83 yr, with a mean of 43.7 yr (Table 40).

The participants differed greatly in how long they had been employed in farm work (including pesticide applications). Because health hazards are proportionate to duration of exposure to toxicants, it was of interest to estimate the ratio of years of employment to age (E/A) for each participant and express the product as percentages. The resultant values may be used in rating degree of occupational exposure

Table 40 Distribution of respondents, by age, involved in field work during a survey study conducted on agricultural laborers at EL-Beheira Governorate, Egypt ($n = 203$ persons)

| Item | Category I[a] <18 yr | Category II 18–39 yr | Category III 40–60 yr | Category IV >60 yr |
|---|---|---|---|---|
| Number | 20 | 60 | 99 | 24 |
| Percent | 9.9 | 29.5 | 48.8 | 11.8 |

[a] Categories are defined in Table 41 (following).

**Table 41** Distribution of respondents by duration (years) of involvement in pesticide field work throughout a survey conducted on agricultural laborers at EL-Beheira Governorate, Egypt ($n = 203$ persons)

| Category (rating in parentheses)[a] | No. of people responding | Response (%) |
|---|---|---|
| I. Excessive Occupational Exposure (>70%) | 31 | 15.3 |
| II. Extreme Occupational Exposure (56%–70%) | 62 | 30.5 |
| III. High Occupational Exposure (36%–55%) | 67 | 33.0 |
| IV. Moderate Occupational Exposure (26 – 35%) | 37 | 18.2 |
| V. Low Occupational Exposure (<20%) | 6 | 3.0 |
| Totals: | 203 | 100.0 |

[a]Values in parentheses are expressed as a percent and calculated by dividing age by the number of years an individual was employed in pesticide-related work.

to pesticides among those in the studied group. In this respect, we proposed five categories of relative exposure intensity (Table 41).

Category I: "Excessive Occupational Exposure" for individuals with an E/A ≥ 70%. These represented 15.3% of the total.

Category II: "Extreme Occupational Exposure" for individuals with E/A = 56%–70%. These represented 30.5% of the total.

Category III: "High Occupational Exposure" for individuals with E/A = 36%–55%. These represented 33.0% of the total.

Category IV: "Moderate Occupational Exposure" for individuals with E/A = 20%–35%. These represented 18.2% of the total.

Category V: "Low Occupational Exposure" for individuals with E/A ≤ 20%. These represented 3.0% of the total.

It was noted that the youngest group (<18 yr) included some workers of 15–17 yr who claimed they had about 10 yr involvement in field pesticide applications, which means they were originally employed at the age of 5–7 yr. According to data presented in Table 41, such workers should be grouped within Category II (Extreme Occupational Exposure). Involvement of young children in farm work, especially those collecting cotton leafworm egg masses and carrying the high-pressure rubber hoses of ground sprayers, is common throughout Egypt during the cotton growing season. The report "Human Rights Watch, Egypt: Underage and Unprotected" (HRW 2006) proposes useful recommendations regarding such "Use of Child Labor in Cotton Pest Management."

Responses of the participants to key questions are presented in Table 42. Among the 203 participants in our questionnaire, 75 (37%) were uneducated (illiterate). Only 8.7% of the total had received a university education. The remaining participants had various levels of education. Nearly half the participants (95 persons, 46.8%) declared that they did not wear protective clothing during pesticide application (see Table 42). Such behavior poses significant health risks to farm workers.

**Table 42** Respondents' answers to key questions in a survey of pesticide field workers at El-Beheira Governorate, Egypt ($n = 203$ persons)

| Question | Response | No. of people responding | Response (%) |
|---|---|---|---|
| Q1: What is your education level? | Incomplete elementary/no schooling | 75 | 37.0 |
| | Elementary | 24 | 11.8 |
| | Preparatory | 46 | 22.8 |
| | Secondary | 40 | 19.7 |
| | Some college | 18 | 8.7 |
| Q2: Do you wear protective clothing? | Never | 95 | 46.8 |
| | Sometimes | 65 | 32.0 |
| | Always | 43 | 21.2 |
| Q3: What is your way of disposing of empty pesticide containers? | Used to store water/grains | 22 | 10.8 |
| | Sell them | 14 | 6.9 |
| | Give to neighbors/friends | 12 | 5.9 |
| | Burn or bury them | 60 | 29.6 |
| | Throw into canals | 80 | 39.4 |
| | Throw into rubbish | 15 | 7.4 |
| Q4: Do you think pesticides leave residues on plants? | Yes | 50 | 24.6 |
| | No | 40 | 19.7 |
| | Not sure | 113 | 55.7 |
| Q5: Who taught you about application of pesticides? | Ministry officials | 43 | 21.2 |
| | Neighbors/friends | 160 | 78.8 |
| Q6: What is your source for selecting suitable pesticides? | Pesticide container label | 15 | 7.4 |
| | Cooperatives | 60 | 29.6 |
| | Ministry officials | 30 | 14.8 |
| | Neighbors/friends | 50v | 24.6 |
| | Pesticide's seller | 48 | 23.6 |
| Q7: Do you eat fruit while working? If so, what do you do? | Eat it without washing | 65 | 32.0 |
| | Wash it in water before eating | 100 | 49.3 |
| | Eat only at break time | 38 | 18.7 |
| Q8: Can we stop using pesticides? | Never | 191 | 94.1 |
| | Possible | 12 | 5.9 |

Participants demonstrated diverse behavior when disposing of empty pesticide containers. The most likely action was to throw empty containers into canals (39.4%). Some farmers used such containers to store water or grain (Table 42).

A total of 153 of 203 persons ignored the fact that the crops they handle are contaminated with pesticide residues (see Table 42). Most participants gained employment in agriculture and information on pesticides from friends or neighbors (78.8% of the total) (Table 42). Participants claimed to have selected pesticides and proper methods for their application based on input from various sources according to their level of education (Table 42). Only a small number (38 persons, 18.7%) of participants preferred not to consume food while working with pesticides (Table 42).

Finally, about 94% of participants believe that it is impossible to stop using pesticides (see Table 42). Among the more educated, some suggested use of alternative control measures, such as biological control and integrated pest management (IPM).

In another location, named Menia El-Kamh, Sharkia Governorate, Egypt, Ibitayo (2006) reported results of a questionnaire administered to 188 farmers who were involved in pesticide field work. Data in Table 43 compare results of certain questions given to Sharkia Group Farmers (SGF) and Beheira Group Farmers (BGF). From the data presented in Table 43, the following can be deduced:

1. Illiteracy is higher in the SGF group, although they are more likely than the BGF group to use protective clothing.
2. Both groups behaved poorly in their handling of empty pesticide containers, although the BGF group was generally better.
3. Both ministry officials and cooperatives seem to have a moderate role in guiding the use and application of pesticides. About 57.0% (SGF) and 44.4% (BGF) of the surveyed populations used these sources to obtain information on pesticide use and application.

Taking into consideration the large number of workers involved in cotton pesticide application and the attitude and behaviors on pesticides they expressed in the foregoing surveys, solutions to the problem of health risks among workers must be further addressed.

## 6 Conclusions

- Once used in an area, persistent pesticides leave residues in various environmental compartments that remain for days to many years.
- OC pesticides, prohibited since the early 1980s, are still detectable in the environment.
- OC pesticides are usually found in the environment and in food at very low residue levels. Such levels are normally within safety margins to human health and are consonant with existing toxicological safety guidelines. As toxicology and guidelines advance, existing residue levels may or may not be regarded to pose health risks.

**Table 43** Comparison of Egyptian farmers' attitudes and behavior on pesticide handling practices in two surveys conducted in different Egyptian provinces

| | Percent response | |
|---|---|---|
| Item | Sharkia Province[1] | Beheira Province[2] |
| Q1: What is your education level? | | |
| Incomplete elementary/no schooling | 54.5 | 37.0 |
| Elementary | 13.9 | 34.6[a] |
| Secondary | 21.9 | 19.7 |
| College | 9.6 | 8.7 |
| Q2: Do you wear protective clothing? | | |
| Never | 33.0 | 46.8 |
| Sometimes | 18.1 | 32.0 |
| Always | 48.9 | 21.2 |
| Q3: Do you think a pesticide leaves residues on plants? | | |
| Yes | 27.8 | 24.6 |
| No | 10.2 | 19.7 |
| Not Sure | 62.0 | 55.7 |
| Q4: How do you dispose of empty pesticide containers? | | |
| Used to store water/grains | 42.4[b] | 10.8[c] |
| Sell them | 23.2 | 6.9 |
| Give to neighbors/friends | 13.0 | 5.9 |
| Throw in wooded areas/rubbish | 7.3 | 7.4 |
| Burn or bury them | 9.0 | 29.6 |
| Throw into canals | 4.0 | 39.4 |
| Q5: What is your source for selecting suitable pesticides? | | |
| Ministry officials | 55.9 | 14.8 |
| Cooperatives | 1.1 | 29.6 |
| Neighbors/friends | 30.2 | 24.6 |
| Pesticide seller | 12.3 | 23.6 |
| The pesticide label | ND | 7.4 |

ND, no data available.
[a] "Basic Education" includes elementary and preparatory schooling.
[b] Used for drinking water for either humans or animals.
[c] Used for either water or grains.
*Sources*: Ibitayo (2006) (1); data from the Beheira survey (2).

- Recent work to identify and measure residues of dioxins and dibenzofurans in the environment does not mean these potent toxicants were absent in the past. Lack of adequate analytical methods and sophisticated equipment may be reasons for not having formerly detected these potent toxicants.
- Comparing residues of a pesticide in a food with standard safety criteria (e.g., MRLs) alone is not sufficient, considering the lack of safety criteria for food contaminated with mixtures of pesticide residues.
- In Egypt, no national strategic programs for monitoring pesticide residues have yet been created. A dearth of such data makes it difficult to fully estimate risks posed by pesticide contamination.

However, in light of existing data, one can say the following:

1. A dietary pesticide intake assessment reveals awarness to certain pesticides that are widely used in Egypt (e.g., dimethoate, dicofol, chlorpyrifos, pirimiphos-methyl).
2. There is inadequate control for pesticide use in Egypt. This is true for both field and indoor uses. In particular, too many underage workers are too frequently involved with agriculture, and laborors who use pesticides are not adequately protected with masks, clothing, or an adequate understanding of the risks to which they are exposed.
3. The recent undertaking of the Egyptian Ministry of Agriculture to reduce numbers and quantities of traditional pesticides, shift use to biocontrol measures, and utilize IPM programs may minimize pesticide contamination and reduce national pesticide risks.

# 7 Summary

The first use of petroleum-derived pesticides in Egyptian agriculture was initiated in 1950. Early applications consisted of distributing insecticidal dusts containing DDT/BHC/S onto cotton fields. This practice was followed by use of toxaphene until 1961. Carbamates, organophosphates, and synthetic pyrethroids were subsequently used, mainly for applications to cotton. In addition to the use of about 1 million metric tons (t) of pesticides in the agricultural sector over a 50-yr period, specific health and environmental problems are documented in this review. Major problems represented and discussed in this review are human poisoning, incidental toxicity to farm animals, insect pest resistance, destruction of beneficial parasites and predators, contamination of food by pesticide residues, and pollution of environmental ecosystems. Several reports reveal that chlorinated hydrocarbon pesticide residues are still detectable in several environmental compartments; however, these residues are in decline. Since 1990, there is a growing movement toward reduced consumption of traditional pesticides and a tendency to expand use of biopesticides, including "Bt," and plant incorporated protectants (PIPs). On the other hand, DDT and lindane were used for indoor and hygienic purposes as early as 1952. Presently, indoor use of pesticides for pest control is widespread in Egypt. Accurate information concerning the types and amounts of Egyptian household pesticide use, or numbers of poisoning or contamination incidents, is unavailable. Generally, use of indoor pesticides is inadequately managed. The results of a survey of Egyptian farmers' attitudes toward pesticides and their behavior in using them garnered new insights as to how pesticides should be better controlled and regulated in Egypt.

**Acknowledgments** The author thanks Prof. Dr. Hany Gamalluldin, Director of Poison Control Center of Ain Shams University Hospitals (PCCA), Cairo, Egypt, for providing statistical data on patterns of poisoning.

# References

Abbassy MS (2000) Pesticides and polychlorinated biphenyls drained into north coast of the Mediterranean Sea, Egypt. Bull Environ Contam Toxicol 64:508–517.

Abbassy MS (2001) Pesticide residues in selected vegetables and fruits in Alexandria City, Egypt, 1997–1998. Bull Environ Contam Toxicol 67:225–232.

Abbassy MS, Ibrahim HZ, Abo-Elamayem MM (1999) Occurrence of pesticides and polychlorinated biphenyls in water of the Nile River at the Estuaries of Rosetta and Damietta Branches, North Delta, Egypt. J Environ Sci Health B34:255–267.

Abbassy MS, Ibrahim HZ, Abdel-Kader HM (2003) Persistent organochlorine pollutants in the aquatic ecosystem of Lake Manzala, Egypt. Bull Environ Contam Toxicol 70:1158–1164.

HealthAbd-Allah AM (1992) Determination of DDTs and PCBs residues in Abu-Quir and El-Max Bays, Alexandria, Egypt. Toxicol Environ Chem 36:89–97.

Abd-Allah AM (1994) Residue levels of organochlorine pollutants compounds in fish from Abu-Quir and Idku Lake, Alexandria, Egypt. Toxicol Environ Chem 44:65–71.

Abd-Allah AM (1999) Organochlorine contaminants in microlayer and subsurface water of Alexandria coast, Egypt. J Assoc Off Anal Chem 82:391–398.

Abd-Allah AM, Abbas MM (1994) Residue levels of organochlorine pollutants in the Alexandria Region, Egypt. Toxicol Environ Chem 41:239–247.

Abd-Allah AM, Ali HA (1994) Residue levels of chlorinated hydrocarbon compounds in fish from El-Max Bay and Marut Lake, Alexandria, Egypt. Toxicol Environ Chem 42:107–114.

Abdel-Halim KY, Salama AK, El-Khateeb EN, Bakry NM (2006) Organophosphorus pollutants (OPP) in aquatic environment at Damietta Governorate, Egypt: Implications for monitoring and biomarker responses. Chemosphere 63:1491–1498.

Abdelmegid LAM, Salem EM (1996) Trends in the pattern of acute poisoning in Alexandria Poison Center in 1994. Proceedings, 3rd Congress on Toxicology in Developing Countries, Cairo, Egypt (19–23 Nov 1995), vol I, pp 237–250.

Abo-Elamayem M, Saad MA, El-Sebae AH (1979) Water pollution with organochlorine pesticides in Egyptian lakes. Proceedings, International Egypt Germinal Seminar on Environmental Protection from Hazardous Pesticides, Alexandria, Egypt (24–29 March, 1979), pp 94–108.

Abou-Arab AAK (1997) Effect of Ras cheese manufacturing on the stability of DDT and its metabolites. Food Chem 59(1):115–119.

Abou-Arab AAK (1999) Effects of processing and storage of dairy products on lindane residues and metabolites. Food Chem 64:467–473.

Abou-Arab AAK, Abou-Donia MA (2001) Pesticide residues in some Egyptian spices and medicinal plants as affected by processing. Food Chem 72:439–445.

Abou-Arab AAK, Gomaa MNE, Badawy A, Naguib K (1995) Distribution of organochlorine pesticides in the Egyptian aquatic ecosystem. Food Chem 54:141–146.

Abou-Arab AAK, Soliman KM, El-Tantawy ME, Ismail BR, Naguib K (1999) Quantity estimation of some contaminants in commonly used medicinal plants in the Egyptian market. Food Chem 67:357–363.

Ahmed MT, Abdel-Hady E, Yossof K, El-Samahy SK (2001a) Residues of polycyclic aromatic hydrocarbons and heavy metals in bread samples collected from Cairo and Ismailia, Egypt. 1 EHS (1):1-5. DOI:http://dx.doi.org/10. 1065/ehs2001.01.001.

Ahmed MT, Loutfy N, Youssof Y, El-Shiekh E, Eissa IA (2001b) Residues of chlorinated hydrocarbons, polycyclic aromatic hydrocarbons and polychlorinated biphenyls in some marine organisms in Lake Temsah, Suez Canal, Egypt. J Aquat Ecosyst Health Manag 4:165–173.

Ahmed MT, Loutfy N, El-Shiekh E (2002) Residue levels of DDE and PCBs in the blood serum of women in the Port Said region of Egypt. J Haz Mater 89(1):41–48.

Allen-Gill SM, Gubala CP, Wilson RW, Landers DH, Wade TL, Sericano JL, Curtis LR (1998) Organochlorine pesticides and polychlorinated biphenyls (PCBs) in sediments and biota from four US Artic Lakes. Arch Environ Contam Toxicol 33:378–387.

Aly OA, Badawy MI (1981) Organochlorine insecticides in selected agricultural areas in Egypt. Proceedings, International Symposium on Management of Industrial Wastewater in Developing Nations, Alexandria, Egypt (28–31 March 1981), pp 273–281.

Aly OA, Badawy MI (1984) Organochlorine residues in fish from the River Nile, Egypt. Bull Environ Contam Toxicol 33:246–252.

Aman IM, Bluthgen A (1997) Occurrence of residues of organochlorine pesticides and poly-chlorinated biphenyls in milk and dairy products from Egypt. Milchwissenschaft 52:394–399.

Amr MM (1990) Health Hazards of Pesticides. Project of Pesticide Intoxication–Egypt, Final Report. Faculty of Medicine, Cairo University, Egypt and International Developmental Research Center (IDRC), Canada.

Amr MM (1999) Pesticide monitoring and its health problems in Egypt, a Third World Country. Toxicol Lett 107:1–13.

Amr MM, Halim ZS (1997) Psychiatric disorders among Egyptian pesticide applicators and formulators. Environ Res 73:193–199.

Assaad R, Roushdy M (1998) Poverty and Poverty Alleviation Strategies in Egypt. A Report Submitted to the Ford Foundation, Cairo, Egypt.

Attia AM (2005) Risk assessment of occupational exposure to pesticides. In: Linkov I, Ramadan AB (eds) Comparative Risk Assessment and Environmental Decision Making. NATO Science Series, Springer, Netherlands, vol 38(3), pp 349–362.

Badawy MI, Wahaab RA (1997) Environmental impact of some chemical pollutants on Lake Manzala. Int J Environ Health Res 7:161–170.

Badawy MI, El-Dib MA, Aly OA (1984) Spill of methyl parathion in the Mediterranean Sea: a case study at Port-Said, Egypt. Bull Environ Contam Toxicol 32:469–477.

Badawy MI, Wahaab RA, Abou Waly HF (1995) Petroleum and chlorinated hydrocarbons in water from Lake Manzala and associated canals. Bull Environ Contam Toxicol 55:258–263.

Barak NAE, Mason CF (1990) Mercury, cadmium and lead in eels and roach. The effects of size, season and locality on metal concentration in flesh and liver. Sci Total Environ 92:249–256.

Barakat AO (2003) Persistent organic pollutants in smoke particles emitted during open burning of municipal solid wastes. Bull Environ Contam Toxicol 70(1):174–181.

Barakat AO (2004) Assessment of persistent toxic substances in the environment of Egypt. Environ Int 30:309–322.

Barakat AO, Kim M, Qian Y, Wade T (2002) Organochlorine pesticides and PCBs residues in sediments of Alexandria Harbor, Egypt. Mar Pollut Bull 44(12):1421–1434.

Chiou CT, Sheng GY, Manes M (2001) A partition-limited model for the plant uptake of organic contaminants from soil and water. Environ Sci Technol 35:1437–1444.

Codex (1993) Codex Alimentarius Commission. Joint FAO/WHO Food Standards Programme.

Dogheim SM, Nasr EN, Almaz MM, El-Tohamy MM (1990) Pesticide residues in milk and fish samples collected from two Egyptian governorates. J Assoc Anal Chem 73:19–21.

Dogheim SM, El-Shafeey M, Afifi AM, Abdel-Aleem FE (1991) Levels of pesticide residues in Egyptian human milk samples and infant dietary intake. J Assoc Off Anal Chem 74(1):89–91.

Dogheim SM, Gad Alla SA, El-Syes SMA, Almaz MM, Salama EY (1996a) Organochlorine and organophosphorus pesticide residues in food from Egyptian local markets. J Assoc Off Anal Chem Int 79(4):949–952.

Dogheim SM, El-Zarka M, Gad Alla SA, Almaz SA, El-Saied S, Salama EY, Mohsen AM (1996b) Monitoring of pesticide residues in human milk, soil, water, and food samples collected from Kafr El-Zayat Governorate. J Assoc Anal Chem 79(1):111–116.

Dogheim SM, Gad Alla SA, El-Marsafy AM (1999) Monitoring pesticide residues in Egyptian fruits and vegetables in 1995. J Assoc Off Anal Chem Int 82(4):948–955.

Dogheim SM, Gad Alla SA, El-Marsafy AM (2001) Monitoring of pesticide residues in Egyptian fruits and vegetables during 1996. J Assoc Off Anal Chem Int 84(2):519–531.

Dogheim SM, El-Marsafy AM, Salama EY, Gad Alla SA, Nabil YM (2002) Monitoring of pesticide residues in Egyptian fruits and vegetables during 1997. Food Addit Contam 19(11):1015–1027.

Dogheim SM, El-Marsafy AM, Gad Alla SA, Khorshid MA, Fahmy SM (2004) Pesticides and heavy metals levels in Egyptian leafy vegetables and some aromatic medicinal plants. Food Addit Contam 21(4):323–330.

Egyptian Organization for Standardization and Quality Control (EOS) (1991) Maximum Residue Limits for Pesticides in Foods. Egyptian Organization for Standardization and Quality Control, Cairo, Egypt.

El-Dib MA, Badawy MI (1985) Organo-chlorine insecticides and PCBs in River Nile water, Egypt. Bull Environ Contam Toxicol 34:126–133.

El-Gamal A (1983) Persistence of some pesticides in semi-arid conditions in Egypt. Proceedings, International Conference on Environmental Hazards in Agrochemistry, Alexandria, Egypt (8–12 Nov, 1983), vol I, pp 54–75.

El-Gendy KS, Abd-Allah AM, Ali HA, Tantawy G, El-Sebaae AE (1991) Residue levels of chlorinated hydrocarbons in water and sediment samples from Nile Branches in the Delta, Egypt. J Environ Sci Health 26:15–36.

El-Hamid A (1993) Pesticide Residues in Human Milk. Ph.D. Thesis, Faculty of Medicine, Menoufiya University, Egypt.

El-Kabbany S, Rashed MM, Zayed MA (2000) Monitoring of the pesticide levels in some water supplies and agricultural land, in El-Haram, Giza (A.R.E.). J Hazard Mater A72:11–21.

El-Kady AA, Abou-Arab AAK, Khairy M, Morsi S, Galal SM (2001) Survey study on the presence of some pesticides and heavy metals in local and imported meat and effect of processing on them. J Egypt Soc Toxicol 24:81–85.

El-Nabawy A, Heinzow B, Kruse H (1987) Residue levels of organochlorine chemicals and polychlorinated biphenyls in fish from the Alexandria region, Egypt. Arch Environ Contam Toxicol 16:689–696.

El-Nemr A, Abd-Allah AMA (2004) Organochlorine contamination in some marketable fish in Egypt. Chemosphere 54:1401–1406.

El-Sebae AH (1977) Incidents of local pesticide hazards and their toxicological interpretation. Proceedings, Seminar/Workshop on Pesticide Management, UC/AID–University of Alexandria, Alexandria, Egypt (5–10 March, 1977), pp 137–152.

El-Sebae AH, Abo-Elamayem M (1978) A survey of expected pollutants drained to the Mediterranean in the Egyptian Region. Proceedings, XXXVI Congress and Plenary Assembly of the International Commission of Scientific Exploration of the Mediterranean Sea, Antalya, Turkey, pp 149–153.

El-Sebae AH, Soliman SA (1982) Mutagenic and carcinogenic chemicals in the Egyptian agricultural environment. Basic Life Sci 21:119–126.

El-Sebae AH, Soliman SA, Ahmed NS (1979) Delayed neurotoxicity in sheep by the phosphorothioate insecticide Cyanophenphos. J Environ Sci Health B14 3:347–362.

El-Sebae AH, Soliman SA, Ahmed NS, Curley A (1981) Biochemical interaction of six OP delayed neurotoxicants with several neurotargets. J Environ Sci Health B 16:463–474.

El-Sebae AH, Abou-Zeid M, Saleh MA (1993) Status and environmental impact of toxaphene in the Third World: a case study of African agriculture. Chemosphere 27(10):2063–2072.

Ezzat S, Abdel-Hamid M, Eissa SA, Mokhtar, N, Labib NA, El-Ghorory L, Mikhail NN, Abdel-Hamid A, Hifnawy T, Strickland GT, Loffredo CA (2005) Association of pesticides, HCV, HBV, and hepatocellular carcinoma in Egypt. Int Hyg Environ Health 208:329–339.

Fahmy S (1998) Monitoring of organochlorine pesticide residues in milk powder. J Drug Res 22:281–291.

FAO/WHO (1993) Food Standards Programme. Pesticide Residues in Food. Volume 2. Codex Alimentarius.

Farahat TM, Abdelrasoul GM, Amr MM, Shebl MM, Farahat FM, Anger WK (2007) Neurobehavioral effects among workers occupationally exposed to organophosphorus pesticides. Occup Environ Med 60:279–286.

Fries GF (1995) A review of the significance of animal food products as potential pathways of human exposures to dioxins. J Anim Sci 73(6):1639–1650.

Garcia AM (1998) Occupational exposure to pesticides and congenital malformations: a review of mechanisms, methods and results. Am J Ind Med 33:232–240.

Hong H, Zhang L, Chen JC, Wong YS, Wan TSM (1995) Environmental fate and chemistry of organic pollutants in the sediment of Xiamen and Victoria Harbors. Mar Pollut Bull 31:229–236.

HRW (2006) Human Rights Watch, Egypt: Underage and Unprotected. http://www.hrw.org/reports/2001/egypt/Egypt01-01.htm.

Ibitayo OO (2006) Egyptian farmers' attitudes and behaviors regarding agricultural pesticides: implications for pesticide risk communication. Risk Anal 26(4):989–995.

Khaled A, El Nemr A, Said TO, El-Sikaily A, Abd-Alla AMA (2004) Polychlorinated biphenyls and chlorinated pesticides in mussels from the Egyptian Red Sea coast. Chemosphere 54:1407–1412.

Loutfy N, Fuerhacker M, Tundo P, Raccanelli S, El Dien AG, Ahmed MT (2006) Dietary intake of dioxins and dioxin-like PCBs, due to the consumption of dairy products, fish/seafood and meat from Ismailia City, Egypt. Sci Total Environ 370(1):1–8.

Loutfy N, Fuerhacker M, Tundo P, Raccanelli S, Ahmed MT (2007) Monitoring of polychlorinated dibenzo-$p$-dioxins and dibenzofurans, dioxin-like PCBs and polycyclic aromatic hydrocarbons in food and feed samples from Ismailia City, Egypt. Chemosphere 66(10):1962–1970.

Mackay D, Leeinonen PJ (1975) Rate of evaporation of low solubility contaminants from water bodies to atmosphere. Environ Sci Technol 9:1178–1180.

Mackay D, Wolkoff AW (1973) Rate of evaporation of low solubility contaminants from water bodies to atmosphere. Environ Sci Technol 7:257–264.

Mackay D, Masearenhas R, Shiu WY, Valvani SC, Yalkowsky SH (1980) Aqueous solubility of polychlorinated biphenyls. Chemosphere 9:178–264.

Mansour SA (1993) Usages and problems of pesticides in Egypt. Int J Toxicol Occup Environ Health 2(2):46–53.

Mansour SA (1998) Perspectives on the use of daphnids in aquatic toxicology of pesticides. In: Majumdar SK, Miller EW, Brenner FJ (eds) Ecology of Wetlands and Associated Systems. The Pennsylvania Academy of Sciences, Easton, pp 372–399.

Mansour SA (2004) Pesticide exposure: Egyptian scene. Toxicology 198:91–115.

Mansour SA (2006) Monitoring of pesticides and heavy metals in the western desert lakes of Egypt and bioassaying potential toxicity of their waters. In: Proceedings, 4th Environmental Conference on Pesticides and Related Organic Micropollutants in the Environment/10th Symposium on the Chemical Fate of Modern Pesticides, University of Almeria, Almeria, Spain, 26–29 Nov, 2006, pp 113–117.

Mansour SA, Sidky MSM (2002) Ecotoxicological studies. 3. Heavy metals contaminating water and fish from Fayoum Governorate, Egypt. Food Chem 78:15–22.

Mansour SA, Sidky MSM (2003) Ecotoxicological studies. 6. The first comparative study between Lake Qarun and Wadi El-Rayan wetland (Egypt), with respect to contamination of their major components. Food Chem 82:181–189.

Mansour SA, Messeha SS, Ibrahim AW (2001a) Ecotoxicological studies. 5. The use of *Daphnia magna* neonates and *Culex pipiens* larvae to bioassay toxicity of Lake Qarun water (Egypt), and to propose preliminary remedial criteria. J Egypt Ger Soc Zool 36(D):115–140.

Mansour SA, Mahran MR, Sidky MSM (2001b) Ecotoxicological studies. 4. Monitoring of pesticide residues in the major components of Lake Qarun, Egypt. J Egypt Acad Soc Environ Dev 2(1):83–116.

Moses M, Johnson ES, Anger WK, Buse VW, Horstman SW, Jackson RJ, Lewis RG, Maddy KT, McConnell R, Meggs WT, Zahm SH (1993) Environmental equity and pesticide exposure. Toxicol Ind Health 9(5):913–959.

Moursy A, Ibrahim MB (1999) Monitoring of organochlorine pollutants in the water of Lake Manzala. WEFTEC "99: 72nd Annual Conference Exposition, Cairo, Egypt.

Mumtaz MM (1995) Risk assessment of chemical mixtures from a public health perspective. Toxicol Lett 82-83:527–532.

PAN (1993) Toxaphene in North Sea Fish. Global Pesticide Campaigner. Pesticide Action Netw 3(4):1–17.

Potter TL, Mohamed MA, Ali H (2007) Solid-phase combined with high-performance liquid chromatography-atmospheric pressure chemical ionization-mass spectrometry analysis of pesticides in water: method performance and application in a reconnaissance survey of residues in drinking water in greater Cairo, Egypt. J Agric Food Chem 55:204–210.

Salim A (2006) Evaluation of heavy metal contents and organochlorine pesticides (OCPs) residues in Egyptian organically-farmed vegetables. J Agric Sci Mansoura Univ 31(3):1601–1612.

Salim A, Zohair A (2004) Effect of acidic natural antioxidant on removal pesticides from some contaminated cereals. J Agric Sci Mansoura Univ 29(8):4675–4683.

Sarkar A, Everaarts JM (1998) Riverine input of chlorinated hydrocarbons in the coastal pollution. In: Majumdar SK, Miller EW, Brenner FJ (eds) Ecology of Wetlands and Associated Systems. The Pennsylvania Academy of Sciences, Easton, PA, pp 400–423.

Sherif SO, Abdulla EA, Morsy AF, Ahmad RT (2005) Chronic exposure to organochlorine pesticides in urban Giza school children. Med J Cairo Univ 73(2):241–255.

Soliman AS, Wang X, DiGiovanni J, Eissa S, Morad M, Vulimiri S, Mahgoub KG, Johnston DA, Do K-A, Seifeldin IA, Boffetta P, Bondy ML (2003) Serum organochlorine levels and history of lactation in Egypt. Environ Res 92:110–117.

Tanabe S, Tatsukawa R, Kawano M, Hidaka H (1982) Global distribution and atmospheric transport of chlorinated hydrocarbons: HCH (BHC) isomers and DDT compounds in the western Pacific, eastern Indian and Antarctic Oceans. J Oceanogr Soc Jpn 38:137–148.

Tricker AR, Preussmann R (1990) Chemical food contaminants in the initiation of cancer. Proc Nutr Soc 49:133–144.

Tundo P, Raccanelli S, Reda L, Ahmed MT(2004) Distribution of polychlorinated dibenzo-$p$-dioxins, polychlorinated dibenzofurans, dioxin-like polychlorinated biphenyls and polycyclic aromatic hydrocarbons in sediment of Temsah Lake, Suez Canal, Egypt. Chem Ecol 20:257–265.

Tundo P, Reda LA, Mosleh YY, Ahmed MT (2005) Residues of PCDD, PCDF, and PCBs in some marine organisms in Lake Temsah, Ismailia, Egypt. Toxicol Environ Chem 87(1):21–30.

UNDP, Egypt (1999) Sustainable Development Documents. Egypt Workshop Report.

Wade TL, Sericano JL, Gardinali PR, Wolff G, Chambers L (1998) NOAA's "Mussel Watch" Project: current use of organic compounds in bivalves. Mar Pollut Bull 37:20–26.

Wahaab RA, Badawy MI (2004) Water quality assessment of the River Nile system: an overview. Biomed Environ Sci 17:87–100.

WHO (1994) Guidelines for Drinking Water Quality. World Health Organization, Geneva.

WHO (2003) GEMS/FOOD REGIONAL DIETS. Food Safety Department, World Health Organization, Geneva, Switzerland. http://www.who.int/foodsafety.

Yamashita N, Urushigawa Y, Masunaga S, Walash M, Miyazaki A (2000) Organochlorine pesticides in water, sediment and fish from the Nile River and Manzala Lake in Egypt. Int J Environ Anal Chem 77:289–303.

Zohair A, Salim A, Soyibo AA, Beck AJ (2006) Residues of polycyclic aromatic hydrocarbons (PAHs), polychlorinated biphenyls (PCBs) and organochlorine pesticides in organically-farmed vegetables. Chemosphere 63:541–553.

# Biodegradation of Perfluorinated Compounds

John R. Parsons, Monica Sáez, Jan Dolfing, and Pim de Voogt

**Contents**

| | | |
|---|---|---|
| 1 | Introduction | 53 |
| 2 | Biodegradation of Organohalogen Compounds | 56 |
| 3 | Thermodynamic Aspects | 58 |
| 4 | Biodegradation of Polyfluorinated Compounds (PFCs) | 60 |
| 5 | Perspectives for the Biodegradation of Perfluorinated Compounds | 64 |
| 6 | Summary | 68 |
| References | | 68 |

## 1 Introduction

In recent years, there has been increasing concern over the levels of perfluorinated and polyfluorinated chemicals, such as PFOS (perfluorosulfonate) and PFOA (perfluorooctanoic acid), in the global environment and their fate and possible adverse effects in the environment. Perfluorinated compounds (PFCs) are substances with the general formula $F(CF_2)_n\text{-}R$, consisting of a hydrophobic alkyl chain of varying length (typically C4 to C16) and a hydrophilic end group (Table 1). The partially fluorinated compounds that contain a $-CH_2CH_2-$ moiety between the hydrophilic part and the fully

---

J.R. Parsons
Institute for Biodiversity and Ecosystem Dynamics, University of Amsterdam,
Nieuwe Achtergracht 166, 1018 WV Amsterdam, The Netherlands

M. Sáez
Department of Instrumental Analysis and Environmental Chemistry, Institute of Organic Chemistry, CSIC, C/ Juan de la Cierva 3, 28006 Madrid, Spain

J. Dolfing
School of Civil Engineering and Geosciences, Newcastle University, Newcastle NE1 7RU, UK

P. de Voogt
Institute for Biodiversity and Ecosystem Dynamics, University of Amsterdam,
Nieuwe Achtergracht 166, 1018 WV Amsterdam,
The Netherlands Kiwa Water Research, Groningenhaven, Nieuwegein, The Netherlands

**Table 1** Examples of perfluorinated compounds

| Name | Chemical formula | Acronym | CAS no.[a] |
|---|---|---|---|
| *Sulfonates* | | | |
| Perfluorobutane sulfonate | $F(CF_2)_4SO_3^-$ | PFBS | 375-73-5 |
| Perfluorooctane sulfonate | $F(CF_2)_8SO_3^-$ | PFOS | 1763-23-1 |
| *Carboxylates* | | | |
| Perfluorobutanoate | $F(CF_2)_3CO_2^-$ | PFB | 375-22-4 |
| Perfluorooctanoate | $F(CF_2)_7CO_2^-$ | PFO | 335-67-1 |
| Perfluorononanoate | $F(CF_2)_8CO_2^-$ | PFN | 375-95-1 |
| Perfluoroundecanoate | $F(CF_2)_{10}CO_2^-$ | PFUn | 2058-94-8 |
| *Alcohols* | | | |
| Perfluorohexyl ethanol | $F(CF_2)_6CH_2CH_2OH$ | 8-2 FTOH | 678-39-7 |
| Perfluorooctyl ethanol | $F(CF_2)_8CH_2CH_2OH$ | 6-2 FTOH | 647-42-7 |
| *Other* | | | |
| 1*H*,1*H*,2*H*,2*H*-Perfluorooctane sulfonate | $F(CF_2)_6CH_2CH_2SO_3^-$ | 6-2 FTS | 27619-97-2 |
| Perfluorooctyl sulfonamide | $F(CF_2)_8SO_3NH_2$ | PFOSA | 754-91-6 |

[a] For acids, the CAS number is for the protonated form, e.g., $F(CF_2)_3COOH$ (=PFBA).

fluorinated remaining carbon chain ($F(CF_2)_n$-$CH_2CH_2$-X) are called telomer substances and derive their name from the telomerization production process. Telomers are suggested to be precursors of some of the PFCs found in the environment.

The hydrophilic moiety of PFCs can be neutral or positively or negatively charged. The resulting compounds are nonionic, cationic, or anionic surface-active agents, respectively, as a consequence of their amphiphilic character. Examples of neutral end groups are $-CH_2CH_2OH$ (fluorotelomer alcohol) and $-SO_3NH_2$ (perfluoroalkyl sulfonamide (e.g., perfluorooctane sulfonamide, PFOSA). Examples of anionic end groups are the carboxylates ($-COO^-$), the sulfonates ($-SO_3^-$), and the phosphates ($-OPO_3^-$). In cationic PFCs, the fluorinated hydrophobic part is attached to another moiety, such as a quaternary ammonium group.

Perfluorinated compounds show high thermal, chemical, and biological inertness. Fluorochemical products can resist degradation by acids, bases, oxidants, reductants, photolytic processes, microbes, and metabolic processes (Schultz et al. 2003), because the carbon–fluorine bond is the strongest existing covalent bond (450 kJ/mol). Moreover, the presence of three pairs of nonbonding electrons around each fluorine atom and the effective shielding of carbon by the fluorine atoms prevents any significant attack (Kissa 2001). Consequently, these fluorosurfactants as well as other fluorinated alkyl substances are stable under conditions in which their hydrocarbon analogues are degraded.

Two major processes exist (Hekster et al. 2003; de Voogt et al. 2006) for production of PFCs, viz., electrochemical fluorination (ECF) and telomerization (TM) (Fig. 1). The products from the first process (the so-called ECF products) contain a

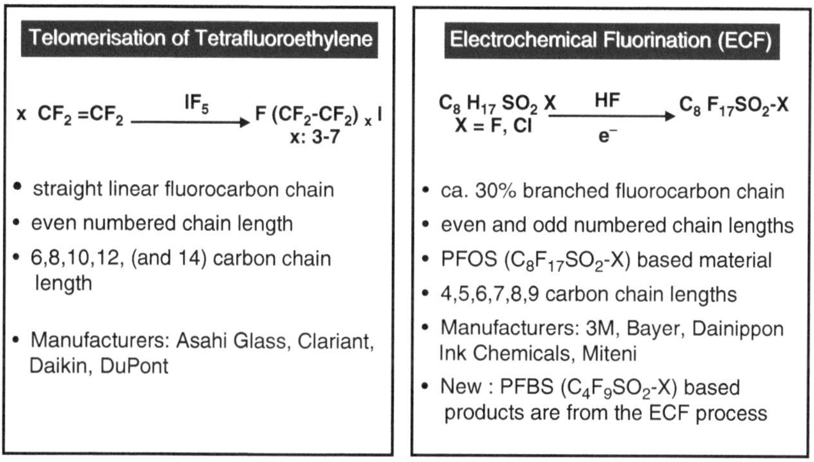

| Telomerisation of Tetrafluoroethylene | Electrochemical Fluorination (ECF) |
|---|---|
| $x\ CF_2=CF_2 \xrightarrow{IF_5} F(CF_2\text{-}CF_2)_x I$ <br> $x: 3\text{-}7$ | $C_8H_{17}SO_2X \xrightarrow[e^-]{HF} C_8F_{17}SO_2\text{-}X$ <br> $X = F, Cl$ |
| • straight linear fluorocarbon chain <br> • even numbered chain length <br> • 6,8,10,12, (and 14) carbon chain length <br><br> • Manufacturers: Asahi Glass, Clariant, Daikin, DuPont | • ca. 30% branched fluorocarbon chain <br> • even and odd numbered chain lengths <br> • PFOS ($C_8F_{17}SO_2$-X) based material <br> • 4,5,6,7,8,9 carbon chain lengths <br> • Manufacturers: 3M, Bayer, Dainippon Ink Chemicals, Miteni <br> • New : PFBS ($C_4F_9SO_2$-X) based products are from the ECF process |

**Fig. 1** The telomerization (*left*) and electrochemical fluorination (*right*) processes for the synthesis of perfluorinated compounds

sulfonyl group. The products from the second production process (telomers) contain an ethylene group. Perfluorooctane sulfonylfluoride (POSF, $C_8F_{17}SO_2F$) is the most important production intermediate for electrochemical fluorination. Fluorotelomer B alcohol 8:2 (8:2-FTOH, $C_8F_{17}C_2C_2H_4OH$) is a pivotal substance for telomer production. The ECF production process yields even- and odd-numbered, branched, as well as straight perfluoroalkyl chains, whereas telomerization only yields even-numbered, linear chains.

PFOS and other perfluorinated compounds are widely used in industrial and consumer applications including stain-resistant coatings for textiles, leather, and carpets, grease-proof coatings for paper products approved for food contact, fire-fighting foams, mining and oil well surfactants, floor polishes, and insecticide formulations (Renner 2001; Prevedouros et al. 2006).

The production of fluorinated polymers may result in unreacted or partially reacted starting materials or intermediates that end up in the final product. These fluorochemical residuals are typically present at a total concentration of less than 1% in the final commercialized products (Olsen et al. 2005), and may include PFHS (perfluorohexanesulfonate), PFOSA, N-MePFOSE (*N*-methylperfluorooctane sulfonamidoethanol), and N-EtPFOSE (*N*-ethylperfluorooctane sulfonamidoethanol) in the case of the ECF process, and perfluorooctanoic acid (PFOA) and other carboxylic acids in the case of the TM process.

Most PFCs hitherto observed in the environment are known to be possible end products resulting from ECF. Recently, however, more information has become available that suggests TM building blocks or end products may also be precursors of perfluorinated carboxylates and other PFCs in the atmosphere (Ellis et al. 2004).

The major global producer using the EF process, with manufacturing plants in North America and Europe, ceased using the EF production process by 2002.

This decision may have been spurred by findings of PFCs (notably some sulfonates and carboxylates) such as PFOS and PFOA in blood of both occupationally exposed persons and the general population and in terrestrial, estuarine, and Arctic ecosystems (Hoff et al. 2003; Hoff et al. 2004; Martin et al. 2004a; Martin et al. 2004b). Consequently, the TM-based production has increased in importance.

PFCs are reported to be widely distributed in the environment (Giesy and Kannan 2001; de Voogt et al. 2006), originating primarily from anthropogenic sources. Until now, it has not been well understood how, and via which routes, these substances are transported into the environment. The limited information on PFOS blood levels in the general population reveals an increase in levels between 1974 and 1989 but no further change up to 2001 (Olsen et al. 2005); geographic differences exist, probably from variability in sources, levels, and exposure patterns to these chemicals in various countries (Kannan et al. 2004). Moreover, compositions of other perfluorochemicals to which the population may be exposed may vary with country and ethnic group within a country (Kannan et al. 2004; Calafat et al. 2006). Information is lacking, however, on which constitute the most important routes of human exposure. Because PFCs are found in environmental biota, it is likely that food is a human exposure route. The relative contribution of the various foodstuffs to the total human exposure is not known.

In general, degradation by microorganisms (biodegradation) is one of the most important mechanisms by which organic contaminants are removed from the environment. The biodegradation pathways and ultimate fate of PFCs are still largely unknown, however. The strength of the carbon–fluorine bond is generally believed to be the main factor in limiting the biodegradability of PFCs. Although microorganisms are able to remove nonfluorinated functional groups, the central question is whether they are able to successfully attack and remove PFC fluorine substituents to achieve mineralization of perfluorinated molecules. The purpose of this review is to summarize the available literature on biodegradation of PFCs and to present their major degradation pathways. In addition, we briefly address current knowledge on the biodegradation of organohalogen compounds in general, the thermodynamic aspects of the biodegradation of PFCs in comparison with other organohalogens, and the evidence supporting microbially catalyzed cleavage of C–F bonds.

## 2 Biodegradation of Organohalogen Compounds

In general, organohalogen compounds are characterized by their high stability in the environment. Nevertheless, it has become evident during the past two decades that many of these compounds are susceptible to microbial degradation. It is therefore recognized that microbial degradation of such compounds, although relatively slow, can contribute significantly to their long-term environmental fate (Dolfing 2003; Fetzner 1998; Janssen et al. 2005; Smidt and de Vos 2004).

All organohalogens are characterized by strong carbon—halogen bonds, leading to kinetic stability. It is important, however, to recognize that dehalogenation is a thermodynamically favorable reaction (Dolfing 2003). Under aerobic conditions, organohalogens are degraded by catabolic pathways similar to those for nonhalogenated analogues (Alexander 1999). Halogenation tends to hinder these reactions, possibly leading to the accumulation of partly degraded metabolites. As a result, aerobic degradation is often a cometabolic process, without apparently yielding useful energy.

A key step in the transformation of halogenated compounds into intermediates in common metabolic pathways is the dehalogenation reaction. A number of dehalogenation mechanisms for both aliphatic and aromatic substrates have been identified (Janssen et al. 2005; Fetzner 1998). These mechanisms include haloacid dehalogenation, halohydrin dehalogenation, haloalkane dehalogenation, and reductive dehalogenation. In a number of cases, the enzymes catalyzing the reactions and the genes encoding these enzymes have been characterized (Janssen et al. 2005; Smidt and de Vos 2004). Such information makes it possible to study how microbes or enzymes have evolved the ability to dehalogenate xenobiotic compounds, which presumably derives from naturally occurring dehalogenation mechanisms, and to identify the presence of dehalogenating activity in the field. There are even a few cases in which information exists on the structure of the active site of the enzyme, which enables computational studies on the mechanism of substrate binding and the subsequent dehalogenation reaction.

Many organohalogen compounds are hydrophobic and therefore tend to accumulate in sediments after they enter the aquatic environment. Other organohalogens such as the chlorinated solvents and perfluorinated surfactants are more water soluble and consequently are groundwater contaminants. As sediment and groundwater environments are predominantly anaerobic, reactions not involving molecular oxygen are required to degrade these compounds. In a number of cases, anaerobic catabolic reactions similar to those occurring aerobically have been identified (Fetzner 1998). These reactions are performed by bacteria presumably using alternative electron acceptors to oxygen, such as nitrate and sulfate. However, of more importance are the reductive reactions leading to dehalogenation. Reductive dehalogenation is thermodynamically favorable under reduced conditions and yields more energy than the reduction of most other electron acceptors (Dolfing 2003; Smidt and de Vos 2004).

The voluminous research done on anaerobic biodegradation of organochlorine compounds has demonstrated that many highly chlorinated compounds that were initially considered to be environmentally persistent do, in fact, undergo reductive dehalogenation at polluted locations. The most extensively studied class of compounds are the chlorinated aliphatic solvents such as tetrachloroethene. Evidence for the reductive dechlorination of chlorinated aliphatic compounds has been found at many contaminated groundwater sites and, in some cases, information is available on the bacteria performing this reaction (Fetzner 1998; Smidt and de Vos 2004). Dehalogenating bacteria have been isolated from a broad range of habitats and characterized using modern methods (Smidt and de Vos 2004). These microorganisms belong to different phyla, and many are able to utilize the energy released by the dehalogenating reactions. Therefore, reductive dechlorination is a form of respiration for these bacteria and is sometimes referred to as chlororespiration or halorespiration (McCarty 1997). Probably the most interesting genus of dechlorinating bacteria is *Dehalococcoides*, because these organisms depend on halorespiration for growth and are able to completely dechlorinate tetrachloroethene. Reductive dehalogenation of aromatic compounds has been demonstrated for, among others, chlorinated benzenes, phenols, benzoates, biphenyls, and dioxins by bacteria in sediments (Dolfing 2003; Fetzner 1998; Smidt and de Vos 2004). A number of reductive dehalogenases and their genes have been characterized (Smidt and de Vos 2004). In principle, knowledge of these enzymes makes it possible to detect dehalogenation potential and activity in anaerobic ecosystems.

Comparatively little information is available on the dehalogenation of brominated, iodinated, and fluorinated compounds. Reductive debromination yields more energy than reductive dechlorination (Dolfing 2003), and this is consistent with the observed reductive debromination of, for example, polybrominated biphenyls (PBBs). In fact, brominated biphenyls have been used to induce the reductive dehalogenation of both PBBs and PCBs (polychlorinated biphenyls) in sediment (Bedard et al. 1998). Recent results indicate that reductive debromination is also possible for other compounds such as polybrominated diphenylethers (Gerecke et al. 2005; Skoczynska et al. 2005).

## 3 Thermodynamic Aspects

Thermodynamics provides a useful tool to evaluate whether organisms can obtain energy for growth from catalyzing a certain reaction. A case in point is the reductive dechlorination of 3-chlorobenzoate. When the first observations of reductive dehalogenation were made, it was tacitly assumed that reductive dechlorination of 3-chlorobenzoate was a fortuitous reaction (Suflita et al. 1982). Subsequent awareness that hydrogen was the source of reducing equivalents for the dechlorination reaction (Dolfing and Tiedje 1986), and that 3-chlorobenzoate therefore acted as the electron acceptor, led to the hypothesis that microorganisms can obtain energy for growth from the use of halogenated compounds as electron acceptors (e.g., hydrogenotrophic methanogenic bacteria can grow with bicarbonate as an electron acceptor). At first, it was difficult to prove this hypothesis because of the fastidiousness of the organisms involved, but thermodynamic calculations indicated that a substantial amount of energy would be available from the dechlorination reaction (Dolfing and Tiedje 1987). With the encouragement of this thermodynamic framework, it was subsequently shown that bacteria can indeed obtain energy for growth from the reductive dechlorination of 3-chlorobenzoate (Dolfing 1990). A wide variety of microorganisms are now known to capably grow from this type of reaction (Smidt and de Vos 2004). Thermodynamics support these observations and indicate that reductive dehalogenation (hydrogenolysis) is exergonic for all classes of chlorinated compounds that have been evaluated, including aromatics such as PCBs, chlorophenols, chlorobenzenes, chlorodioxins, and aliphatics such as chloromethanes, chloroethanes, and chlorethenes (Dolfing and Harrison 1992; Dolfing and Janssen 1994; Holmes et al. 1993; Huang et al. 1996).

The amount of energy available from reductive dechlorination varies between 100 and 180 kJ/mol (standard conditions, hydrogen as electron donor) (Dolfing 2003). This amount is much more than that available from sulfate reduction and methanogenesis (38 and 31 kJ/mol $H_2$, respectively), not only under standard conditions, but also under environmentally relevant conditions, i.e., at low hydrogen concentrations. This fact explains why dechlorinating organisms are able to function at hydrogen concentrations too low to sustain methanogenesis (Löffler et al. 1999).

The amount of energy available from dechlorination depends somewhat on the position of the substituents and the presence (and nature) of neighboring groups, although differences between different congeners are generally small (Dolfing 2003). Attempts have been made to rationalize degradation pathways for polychlorinated compounds by assuming that higher energy-yielding steps would take preference over those with a lower yield (Dolfing and Harrison 1993). This approach has seen mixed success (Masunaga et al. 1996), possibly because estimates of the Gibbs free energy of formation values on which these predictions are based are not accurate enough to allow such predictions. However, other explanations can also be envisaged; a major one is that the kinetics of the dehalogenating organisms is also governed by structural and evolutionary factors.

A wealth of data have become available during the past two decades on the Gibbs free energy values of chlorinated compounds. Although less information is available on Gibbs free energy values of fluorinated compounds, available data are clear in showing that the amount of energy available from defluorination is similar to (or slightly lower than) the amount available from dechlorination, as is illustrated in Table 2 for reductive dehalogenation of halobenzoates and halomethanes. The higher energy from dechlorination compared to defluorination is in line with the observation that chlorofluoroalkanes are dechlorinated rather than defluorinated (Hageman et al. 2001; Sonier et al. 1994). However, the substantial energy yield from defluorination also indicates that, from a thermodynamic point of view,

**Table 2** Gibbs free energy values for reductive dehalogenation (hydrogenolysis) of selected fluorinated aromatic and aliphatic compounds and their chlorinated analogues[a]

| | | $\Delta G°$, (kJ/mol) | |
|---|---|---|---|
| | | Defluorination | Dechlorination |
| 2-Halobenzoate + $H_2$ | → Benzoate + $H^+$ + halide⁻ | −132 | −145 |
| 3-Halobenzoate + $H_2$ | → Benzoate + $H^+$ + halide⁻ | −138 | −137 |
| 4-Halobenzoate + $H_2$ | → Benzoate + $H^+$ + halide⁻ | −142 | −144 |
| Halomethane | → Methane + $H^+$ + halide⁻ | −156 | −164 |
| Dihalomethane | → Halomethane + $H^+$ + halide⁻ | −107 | −161 |
| Trihalomethane | → Dihalomethane + $H^+$ + halide⁻ | −85 | −170 |
| Tetrahalomethane | → Trihalomethane + $H^+$ + halide⁻ | −89 | −188 |

[a] All calculations used the following standard conditions: T = 298.15 K, pH 7, methanes and $H_2$ in the gas phase at 1 atm, benzoates and halides in the aqueous phase at 1 M.
*Sources*: Dolfing and Harrison (1992); Dolfing (2003).

there is no reason why microorganisms should not be able to obtain energy for growth from reductive defluorination. So far, however, such organisms have not been described (Vargas et al. 2000).

In the foregoing paragraphs, the emphasis is on whether organisms can obtain energy for growth from catalyzing a certain reaction. It should be remembered that microorganisms do not always harness the energy that their activities generate. Reductive dechlorination is stimulated by the presence of, for example, vitamin $B_{12}$ (Assaf-Anid et al. 1992). Methanogenic bacteria possessing vitamin $B_{12}$ catalyze this reaction, but their growth is not stimulated by the energy released (van Eekert et al. 1999). In fact, the dechlorination reaction costs the organisms energy in the form of reducing equivalents that could have been used to generate energy via methanogenesis.

Cometabolic degradation of halogenated compounds by aerobic microorganisms has also attracted a fair amount of attention (Semprini 1997) and is the basis for technology that has been applied in the field for the degradation of trichloroethene, for example (Hopkins and McCarty 1995). However, from a thermodynamic point of view, there is no reason why halogenated compounds such as tri- and di-chlorethene should be degraded cometabolically rather than serve as an energy source. A change in Gibbs free energy indicates that aerobic mineralization of halogenated compounds yields enough energy to sustain growth of the organism involved (Dolfing et al. 1993; Dolfing and Janssen 1994). Whether organisms harness this energy is another question. Nevertheless, in one case, the fact that thermodynamic calculations support the hypothesis that organisms should be able to grow on halogenated compounds assisted in the isolation of a novel dichloroethene-degrading aerobic bacterium (Coleman et al. 2002).

## 4 Biodegradation of Polyfluorinated Compounds (PFCs)

Despite their high stability, polyfluorinated compounds do undergo degradation in the environment. Atmospheric degradation of fluorotelomer alcohols by OH radicals yields perfluorinated carboxylic acids (Ellis et al. 2004) and has been proposed as a source of the acids present in the Arctic. The acids do react further with OH radicals, but this is considered to be of minor importance (Hurley et al. 2004). The hydroxyl radical also reacts with 8:2 fluorotelomer alcohol in aqueous systems, including Lake Ontario water samples irradiated with synthetic sunlight (Gauthier and Mabury 2005). There was no evidence for direct photolysis of the fluorotelomer alcohol. The main products formed were the 8:2 fluorotelomer aldehyde and acid and PFOA.

The published information on the biodegradation of PFCs is very limited. The earliest study reported was an investigation of the ready biodegradability of three fluorinated surfactants under aerobic and anaerobic conditions (Remde and Debus 1996). Two of the surfactants (of unspecified composition) were degraded under aerobic conditions and one was also degraded anaerobically. The extent of degradation of the first surfactant, expressed as oxygen consumption or carbon dioxide or methane produced, was enough to classify the surfactant as readily biodegradable. However, there was no evidence for the release of fluoride from the compound. No biodegradation was observed for the third surfactant, PFOS, under either aerobic or anaerobic conditions.

An extensive analytical study of the biodegradation of PFCs was reported by Schröder (2003). Wastewater samples were spiked with a number of perfluorinated surfactants (PFOS, PFOA, and nonionic surfactants including partially fluorinated alkyl ethoxylates, perfluorooctanesulfonyl-amidopolyethoxylate, and perfluorooctanesulfonyl-amido–polyethoxylate methyl ether) and incubated under aerobic and anaerobic conditions. Rapid biodegradation was observed in aerobic wastewater of the partly fluorinated compounds to form carboxylic acids (identified by LC-MS/MS). For the perfluorinated compounds, in contrast, there was a rapid removal of PFOS (within 2 d) under anaerobic conditions followed by a slower removal of PFOA. Of the nonionics, only the sulfonyl compounds were removed. Metabolites were neither detected in the anaerobic incubations nor was there any increase in fluoride concentration observed (Schröder 2003).

A further study of the biodegradation of PFOS and PFOA was performed in aerobic and anaerobic reactors containing sludge from German wastewater treatment plants (WWTPs) (Meesters and Schröder 2004). No primary biodegradation was observed under aerobic conditions, but removal, first of PFOS and subsequently of PFOA, was observed under anaerobic conditions until neither compound could be further detected after 26 d. No metabolites or increases in fluoride ion concentration were detected.

The first detailed report providing evidence for environmental biodegradation of PFCs was a study of the biodegradation of 8:2 fluorotelomer alcohol by a mixed microbial consortium enriched from sediment and soil, using 1,2-dichloroethane and ethanol as carbon sources (Dinglasan et al. 2004). Cell suspensions were spiked with 8:2 FTOH and the aerobic incubation was followed by headspace analysis using GC-ECD and GC-MS. FTOH concentrations declined to undetectable levels over a period of about 17 d. Analysis of the aqueous phase by LC-MS/MS revealed formation of acid metabolites: 8:2 fluorotelomercarboxylic acid (FTCA), 8:2 fluorotelomer unsaturated carboxylic acid (FTUCA), and PFOA as major products. The proposed pathway consisted of the oxidation of FTOH to FTCA and the formation of FTUCA from FTCA, presumably by loss of HF and subsequent conversion of FTUCA to PFOA.

Another extensive biodegradation study with 8:2 fluorotelomer alcohol was carried out by Wang et al. (2005a,b). Microbes in a diluted sewage sludge from a domestic WWTP degraded FTOH to the 8:2 FTCA, the 8:2 FTUCA, and PFOA, consistent with the data reported by Dinglasan et al. (2004). The authors also identified a new transformation product, $2H,2H,3H,3H$-perfluorododecanoic acid ($CF_3(CF_2)_6 CH_2CH_2COOH$, also referred to as 7-3 acid), which is a potential substrate for β-oxidation in the degradation pathway (Fig. 2). It was proposed that this compound was formed from the saturated acid by reductive defluorination. Several minor products were identified: 7:2 FTOH, 7:3 FTUCA, and 7:3 fluorotelomer unsaturated amide, as well as PFNA. The release of fluoride to the medium was also significant (about 12% of that present in the telomer alcohol). These results demonstrate that perfluorinated carbon atoms in 8:2 FTOH are indeed defluorinated and the products are degraded by microorganisms from WWTPs to form shorter chain products.

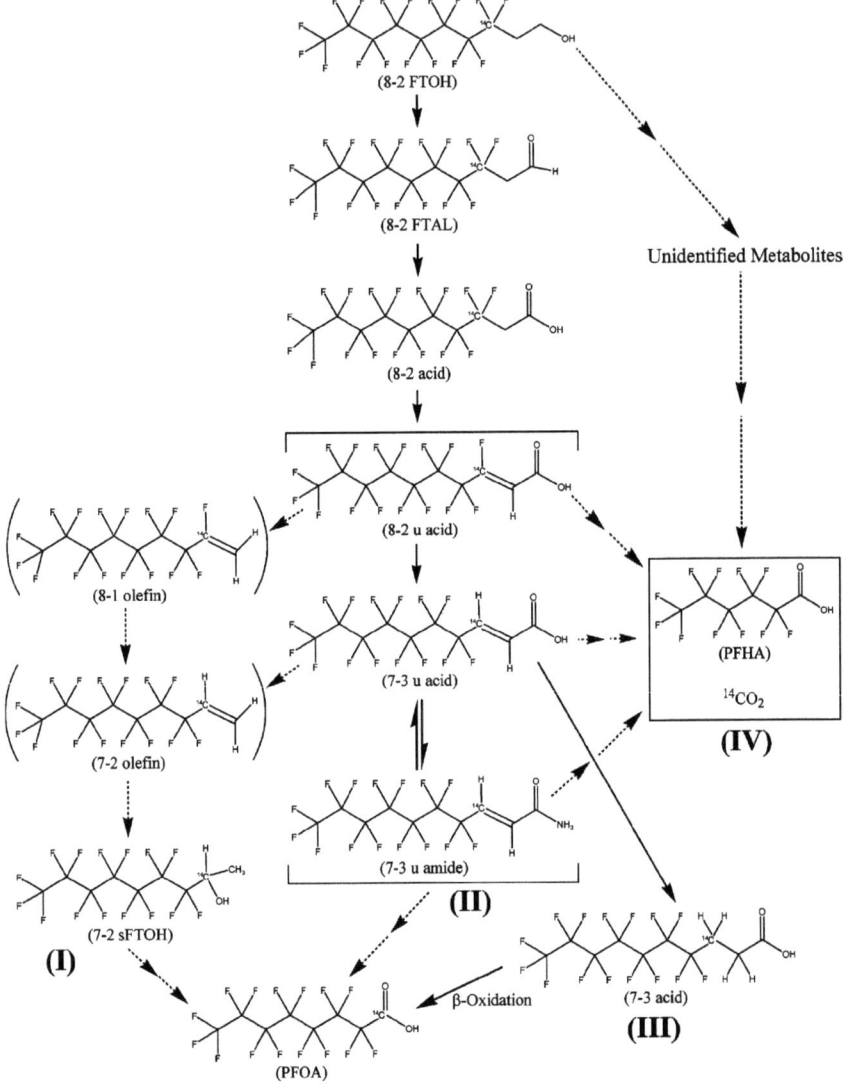

**Fig. 2** Proposed pathway for the degradation of 8-2 fluorotelomer alcohol to perfluorooctanoic acid (reproduced with permission from Wang et al. 2005b)

The formation of PFOA has recently been reported during the aerobic incubation of soil with a fluorotelomer-based polymer. The formation was attributed to the degradation of residual 8:2 FTOH released from the polymer. The half-life of the 8:2 FTOH was estimated as 28 d (Koch et al. 2007).

A preliminary study by the 3M company showed that 2(N-ethyl perfluorooctane sulfonamido)ethanol (N-EtFOSE, used to manufacture polymer surface protection products) was degraded in aerobic waste water treatment sludge (3M 2000). Transformation to PFOS and other metabolites occurred within 25 d, and this compound was not degraded further. A subsequent study of the biodegradation of N-EtFOSE in anaerobic and aerobic sludge showed no degradation in anaerobic sludge but 60% removal in aerobic sludge (Boulanger et al. 2005). The main products identified were 2-(*N*-ethyl-perfluorooctanesulfonamido)acetic acid (N-EtFOSAA) and perfluorooctanesulfinate (PFOsulfinate), accounting for about 23% and 5.3% of the transformed compound, respectively. The remaining transformation products were not identified but did not include the 2-(perfluorooactanesulfonamido)acetic acid (PFOSAA), perfluorooctanesulfonylamide (FOSA), and PFOS metabolites identified in the 3M study. Recent results do, however, show the rapid degradation of N-EtFOSE in activated sludge (with a half-life of 0.25 d) to form N-EtFOSAA and other products including FOSA, PFOS, and possibly even PFOA (Fig. 3) (Rhoads et al. 2007). Formation of the latter product would indicate that defluorination must take place.

A recent study of the degradability of a wide range of PFCs in aerobic and anaerobic sediment and activated sludge microcosms showed some evidence for the removal of perfluorinated telomer alcohols under aerobic conditions (Sáez et al., in preparation). There was, however, no evidence of these compounds in anaerobic

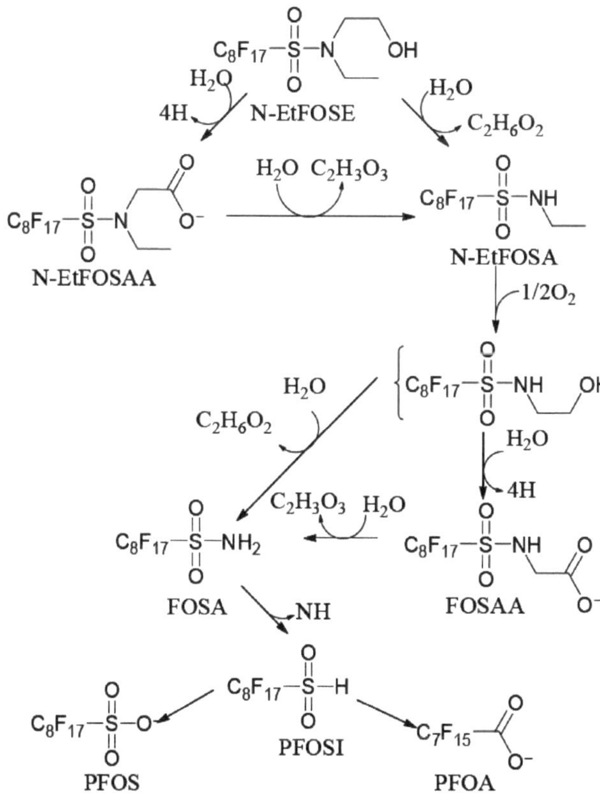

**Fig. 3** Proposed pathway for the degradation of 2(*N*-ethyl perfluorooctanesulfonamido)ethanol (N-EtFOSE) in aerobic activated sludge (Rhoads et al. 2007)

incubations or of other PFCs, including carboxylates, sulfonates, and sulfonamides, under either aerobic or anaerobic conditions.

The foregoing literature shows that there is some evidence for biodegradation of PFCs in the environment, but none for complete degradation or dehalogenation of these compounds, probably because of the difficulty of removing fluorine substituents. However, as previously explained, this is a thermodynamically favorable reaction, but one that may be hindered by the kinetic stability of the C–F bond. One could conclude from these results that the C–F bond is microbially inert. Reductive dehalogenation is a well-established reaction for chlorinated and to a lesser extent brominated organic compounds, particularly under anaerobic conditions, but defluorination may be very difficult to accomplish. For example, dechlorination but not defluorination takes place in sewage sludge with mixed chlorinated and fluorinated substrates (Balsiger et al. 2005; Hageman et al. 2001; Sonier et al. 1994).

There are, however, examples of microbially catalyzed C–F cleavage reactions such as the defluorination reported by Wang et al. (2005a,b). An overview of microbial defluorination reactions was published recently (Natarajan et al. 2005). Examples of known microbially catalyzed defluorination reactions include the oxidation of fluoromethane to formaldehyde by a *Nitrosomonas* strain (Hyman et al. 1994) and the degradation of fluorinated acetates by different aerobic bacteria (Marion Meyer et al. 1990). Similar reactions have been observed for other nonperfluorinated compounds such as the defluorination of 4-fluorobenzoate by six bacterial strains able to grow on this compound (Oltmanns et al. 1989) and the liberation of fluoride from fluorobenzoates and fluorobenzene by another aerobic strain (Carvalho et al. 2005). Defluorination of a number of sulfonates has also been observed (Key et al. 1998). A *Pseudomonas* strain defluorinated several nonperfluorinated compounds, such as $1H,1H,2H,2H$-perfluorooctane sulfonate, but not PFOS. Although the products of these microbial defluorination reactions have not always been characterized, the mechanism involved appears to be either hydrolytic or oxidative. There have been no reports to date of reductive defluorination, despite the fact that this reaction is thermodynamically favorable under reducing conditions. The lack of data on anaerobic reductive defluorination reactions may be a reflection of the very limited studies that have focused on anaerobic conditions. More studies, using anaerobic microbial communities with a history of exposure to fluorinated chemicals, are required.

## 5 Perspectives for the Biodegradation of Perfluorinated Compounds

As already mentioned, thermodynamic calculations indicate that reductive defluorination of fluorinated compounds yields sufficient energy for growth of the organisms catalyzing this type of reaction. Estimated amounts of energy range between

80 and 160 kJ/mol fluoride, assuming equimolar concentrations of $AF_n$ and $AHF_{n-1}$ and environmentally realistic conditions (pH 7; $F^-$ = 0.1 mM, and $H_2$ = 10 Pa). We do not expect these values to be much different for perfluorinated compounds. This hypothesis is based on lines of evidence presented next.

Yamada et al. (1999) published the thermodynamic properties of a series of mono- to pentafluorinated propanes (Table 3). Analysis (data not shown) of the relationship between the number of fluorosubstituents and $T\Delta S$ values of these compound allowed us to also estimate $S°$ for perfluoropropane (i.e., octafluoropropane) and heptafluoropropane (Table 3). In combination with $\Delta H_f°$ values published elsewhere (Kondo et al. 2002), this also allowed us to calculate $G_f°$ values for octa- and heptafluoroethane. With this information, it is straightforward to estimate the change in Gibbs free energy from reductive defluorination of perfluoropropane to heptafluoropropane: 119 kJ/mol (Fig. 4). This value is similar to the amount achieved from reductive removal of other fluorine substituents in the fluoropropane molecule (Table 4). When we initiated these calculations, we anticipated that the amount of energy available from hydrogenolysis of perfluorinated compounds would be less than the amount available from less fluorinated congeners. However, this turned out not to be the case (Fig. 4). We have made a similar comparison for fluorinated ethanes (Table 5, Fig. 5), and the outcome was essentially the same as for fluorinated propanes.

These calculations imply that the energy yield from hydrogenolysis of perfluorinated compounds is not significantly different than that of hydrogenolysis from less fluorinated compounds and suggests that it is reasonable to pose that reductive defluorination of fluorinated compounds, including the perhalogenated ones, is an energy-yielding reaction. Thus, and despite a lack of published thermodynamic

**Table 3** Thermodynamic properties of fluorinated propanes[a]

| Compound[b] | $\Delta H_f°$ kJ/mol | $S°$ J/mol K | $G_f°$ kJ/mol |
|---|---|---|---|
| $CH_2FCH_2CH_3$ | −281.9 | 307 | −201.7 |
| $CHF_2CH_2CH_3$ | −517.4 | 324 | −431.6 |
| $CF_3CH_2CH_3$ | −766.0 | 330 | −671.1 |
| $CH_2FCHFCH_3$ | −459.2 | 325 | −373.7 |
| $CHF_2CHFCH_3$ | −689.0 | 336 | −596.1 |
| $CHF_2CF_2CH_3$ | −905.3 | 360 | −808.9 |
| $CF_3CHFCH_3$ | −927.0 | 346 | −826.2 |
| $CF_3CF_2CH_3$ | −1134.4 | 360 | −1027.1 |
| $CF_3CHFCF_3$ | −1572[c] | 390[d] | −1452.2 |
| $CF_3CF_2CF_3$ | −1178[c] | 404[d] | −1651.5 |

[a] Properties are for ideal gases at a pressure of 1 atm and a temperature of 298.15 K.
[b] Unless stated otherwise, $\Delta H_f$ and $S°$ values are from Yamada et al. (1999).
[c] Data from Kondo et al. (2002).
[d] Calculated based on a correlation between $T\Delta S$ and the number of fluorine substituents ($n$) in fluoropropane ($T\Delta S = 6.7 \times n + 72.9$; $r^2 = 0.97$) in the data series from Yamada et al. (1999).

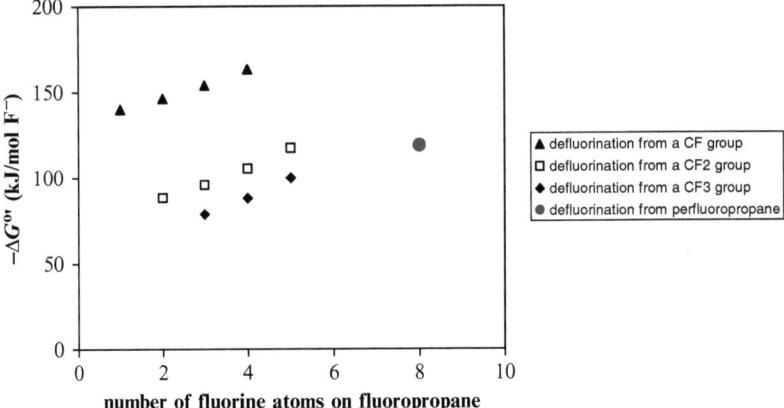

**Fig. 4** Change in Gibbs free energy ($\Delta G^{\circ\prime}$) values for reductive defluorination (hydrogenolysis) of fluorinated propanes

**Table 4** Gibbs free energy changes for reductive defluorination (hydrogenolysis with $H_2$ as electron donor) of fluorinated propanes[a]

| | | | | | Defluorination at position: | | |
|---|---|---|---|---|---|---|---|
| Substrate | Product | $\Delta G^\circ$, (kJ/mol) | | n | $CF_3$ | $CF_2$ | CF |
| $CF_3CF_2CH_3$ | $CHF_2CF_2CH_3$ | −100.0 | | 5 | $F_3F_2 \to F_2F_2$ | | |
| $CF_3CF_2CH_3$ | $CF_3CHFCH_3$ | | −117.4 | 5 | | $F_3F_2 \to F_3F$ | |
| $CF_3CHFCH_3$ | $CHF_2CHFCH_3$ | −88.2 | | 4 | $F_3F \to F_2F$ | | |
| $CHF_2CF_2CH_3$ | $CHF_2CHFCH_3$ | −105.5 | | 4 | | $F_2F_2 \to F_2F$ | |
| $CF_3CHFCH_3$ | $CF_3CH_2CH_3$ | | −163.2 | 4 | | | $F_3F \to F_3$ |
| $CF_3CH_2CH_3$ | $CHF_2CH_2CH_3$ | −78.8 | | 3 | $F_3 \to F_2$ | | |
| $CHF_2CHFCH_3$ | $CH_2FCHFCH_3$ | −96.0 | | 3 | | $F_2F \to FF$ | |
| $CHF_2CHFCH_3$ | $CHF_2CH_2CH_3$ | | −153.9 | 3 | | | $F_2F \to F_2$ |
| $CHF_2CH_2CH_3$ | $CFH_2CH_2CH_3$ | −88.4 | | 2 | | $F_2 \to F$ | |
| $CH_2FCHFCH_3$ | $CFH_2CH_2CH_3$ | | −146.3 | 2 | | | $FF \to F$ |
| $CFH_2CH_2CH_3$ | $CH_3CH_2CH_3$ | | −140.0 | 1 | | | $F \to$ |

[a] $\Delta G^{\circ\prime}$ calculations are for ideal gas-phase defluorination based on data presented in Table 2, with $G_f^\circ$ F⁻ (aq.) = −278.8 kJ/mol and pH =7; n = number of fluorine substituents present on the parent molecule.

properties ($G_f^\circ$, $\Delta H_f^\circ$, $S^\circ$ values) for perfluoroalkylated substances, it seems reasonable to also predict that, thermodynamically, there is no reason why these compounds should not be biodegradable under anaerobic conditions. The question remains, of course, how fast microorganisms will evolve that are able to benefit from this potential source of energy and from where and how they should recruit the enzymatic machinery necessary to catalyze this reaction and harness the energy produced.

**Table 5** Gibbs free energy changes for hydrogenolysis and dihaloelimination of fluorinated ethanes with $H_2$ as electron donor[a]

| Substrate | $\Delta G^{\circ'}$ (kJ/mol) Hydrogenolysis | Difluoroelimination |
|---|---|---|
| $CF_3CF_3$ | −91.0 | −2.9 |
| $CHF_2CF_3$ | −108.6 | −71.2 |
| $CH_2FCF_3$ | −133.4 | −122.4 |
| $CHF_2CHF_2$ | −109.6 | −94.9 |
| $CH_3CF_3$ | −75.7 | na |
| $CH_2FCHF_2$ | −126.2 | −157.6 |
| $CH_3CHF_2$ | −91.5 | na |
| $CH_2FCH_2F$ | −137.5 | −178.0 |
| $CH_3CH_2F$ | −109.0 | Na |

na, not applicable.

[a]$\Delta G^{\circ'}$ calculations are for ideal gas-phase dehalogenation based on data presented in Burgess et al. (1996), Yamada et al. (1998), and Speight (2005) ($G_f^{\circ}$ F⁻ (aq.) ä −278.8 kJ/mol, pH ä7)

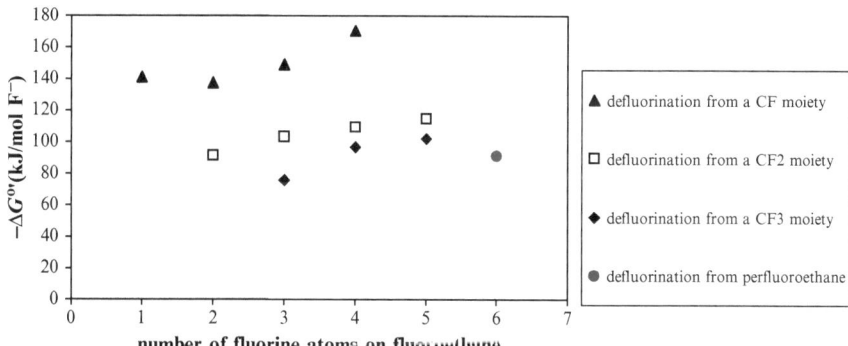

**Fig. 5** Change in Gibbs free energy ($\Delta G^{\circ'}$) values for reductive defluorination (hydrogenolysis) of fluorinated ethanes (based on data from Yamada et al. 1998 and Speight 2005)

Finally, we expect that the organisms that thrive on anaerobic degradation of perfluorinated compounds will perform a hydrogenolytic defluorination rather than dihaloelimination, because hydrogenolysis of, for example, fluorinated ethanes yields more energy than difluoroelimination. This is different for perchlorinated ethanes, where dihaloelimination yields substantially more energy than hydrogenolysis (Dolfing 2000).

In the preceding paragraphs we have focused on reductive defluorination rather than on aerobic degradation of perfluorinated compounds, largely because perhalogenated compounds are generally only degraded under anaerobic conditions (Bossert et al. 2003). Recently, however, an exception to this rule has been created in the laboratory (Yan et al. 2006). Thus, there are no thermodynamic reasons why perfluorinated compounds such as hexafluoroethane or octafluoropropane should not be degradable under aerobic conditions, given that the Gibbs free energy yield ($\Delta G^{\circ'}$) for mineralization of these compounds is estimated to be 730 and 1132 kJ/mol, respectively.

## 6 Summary

The information available in the literature provides evidence for the biodegradation of some poly- and per-fluorinated compounds, but such biodegradation is incomplete and may not result in mineralization. Recent publications have demonstrated that 8:2 fluorotelomer alcohol, for example, can be degraded by bacteria from soil and wastewater treatment plants to perfluorooctanoic acid. Similarly, 2-$N$-ethyl(perfluorooctane sulfonamido)ethanol can be degraded by wastewater treatment sludge to perfluorooctanesulfonate. It is presently unclear whether these two products are degraded further. Therefore, the question remains as to whether there is a potential for defluorination and biodegradation of PFCs that contributes significantly to their environmental fate.

The lack of mineralization observed is probably caused by the stability of the C–F bond, although there are examples of microbially catalyzed defluorination reactions. As is the case with reductive dechlorination or debromination, reductive defluorination is energetically favorable under anaerobic conditions and releases more energy than that available from sulfate reduction or methanogenesis. Consequently, we should consider the possibility that bacteria will adapt to utilize this source of energy, although evolving mechanisms to overcome the kinetic barriers to degradation of these compounds may take some time. The fact that such reactions are absent for some PFCs, to date, may be because too little time has passed for microorganisms to adapt to these potential substrates. Hence, the situation may be comparable to that of chlorinated organic compounds several decades ago. For many years, organochlorine compounds were considered to be catabolically recalcitrant; today, reductive chlorination reactions of many organochlorines, including PCBs and dioxins, are regularly observed in anaerobic environments. Hence, it is opportune and important to continue studying the potential degradation of perfluorinated compounds in carefully designed experiments with either microbial populations from contaminated sites or cultures of bacteria known to dehalogenate chlorinated compounds.

## References

3M (2000) Docket AR-226. U.S. Environmental Protection Agency, Office of Pollution Prevention and Toxic Substances, Washington, DC.
Alexander M (1999) Biodegradation and Bioremediation, 2nd Ed. Academic Press, London.
Assaf-Anid N, Nies L, Vogel TM (1992) Reductive dechlorination of a polychlorinated biphenyl congener and hexachlorobenzene by vitamin $B_{12}$. Appl Environ Microbiol 58:1057–1060.
Balsiger C, Holliger C, Höhener P (2005) Reductive dechlorination of chlorofluorocarbons and hydrochlorofluorocarbons in sewage sludge and aquifer sediment microcosms. Chemosphere 61:361–373.

Bedard DL, van Dort H, DeWeerd KA (1998) Brominated biphenyls prime extensive microbial reductive dehalogenation of Aroclor 1260 in Housatonic River sediment. Appl Environ Microbiol 64:1786–1795.

Bossert I, Haggblom MM, Young LY (2003) Microbial ecology of dehalogenation. In: Häggblom MM, Bossert ID (eds) Environmental Dehalogenation. Kluwer, Dordrecht, pp 33–52.

Boulanger B, Vargao JD, Schnoor, JL, Hornbuckle KC (2005) Evaluation of perfluorooactane surfactants in a wastewater treatment system and in a commercial surface protection product. Environ Sci Technol 38:5524–5530.

Burgess DR Jr, Zachariah MR, Tsang W, Westmoreland PR (1996) Thermochemical and chemical kinetic data for fluorinated hydrocarbons. Prog Energy Combust Sci 21:453–529.

Calafat AM, Needham LL, Kuklenyik Z, Reidy JA, Tully JS, Aquilar-Villalobos M, Naeher LP (2006) Perfluorinated chemicals in selected residents of the American continent. Chemosphere 63:490–496.

Carvalho MF, Ferreira Jorge R, Pacheco CC, De Marco P, Castro PML (2005) Isolation and properties of a pure bacterial strain capable of fluorobenzene degradation as sole carbon and energy source. Environ Microbiol 7:294–298.

Coleman NV, Mattes TE, Gossett JM, Spain JC (2002) Biodegradation of *cis*-dichloroethene as the sole carbon source by a beta-proteobacterium. Appl Environ Microbiol 68:2726–2730.

de Voogt P, Berger U, de Coen, W, de Wolf W, Heimstad E, McLachlan M, van Leeuwen S, van Roon A (2006) Perfluorinated organic compounds in the European environment (Perforce). Report to the EU. University of Amsterdam, Amsterdam, The Netherlands, pp 1–126.

Dinglasan MJA, Ye Y, Edwards EA, Mabury SA (2004) Fluorotelomer alcohol biodegradation yields poly- and perfluorinated acids. Environ Sci Technol 38:2857–2864.

Dolfing J (1990) Reductive dechlorination is coupled to ATP production and growth in an anaerobic bacterium, strain DCB-1. Arch Microbiol 153:264–266.

Dolfing J (2000) Energetics of anaerobic degradation pathways of chlorinated aliphatic compounds. Microb Ecol 40:2–7.

Dolfing J (2003) Thermodynamic considerations for dehalogenation. In: Häggblom MM, Bossert ID (eds) Environmental Dehalogenation. Kluwer, Dordrecht, pp 89–114.

Dolfing J, Harrison BK (1992) Gibbs free energy of formation of halogenated aromatic compounds and their potential role as electron acceptors in anaerobic environments. Environ Sci Technol 26:2213–2218.

Dolfing J, Harrison BK (1993) Redox and reduction potentials as parameters to predict the degradation pathway of chlorinated benzenes in anaerobic environments. FEMS Microbiol Ecol 13:23–29.

Dolfing J, Janssen DB (1994) Estimates of Gibbs free energy of formation of chlorinated aliphatic compounds. Biodegradation 5:21–28.

Dolfing J, Tiedje JM (1986) Hydrogen cycling in a three-tiered food web growing on the methanogenic conversion of 3-chlorobenzoate. FEMS Microbiol Ecol 28:293–298.

Dolfing J, Tiedje JM (1987) Growth yield increase linked to reductive dechlorination in a defined 3-chlorobenzoate degrading methanogenic coculture. Arch Microbiol 149:102–105.

Dolfing J, van den Wijngaard AJ, Janssen DB (1993) Microbiological aspects of the removal of chlorinated compounds from air. Biodegradation 4:261–282.

Ellis DA, Martin JW, De Silva AO, Mabury SA, Hurley MD, Andersen MPS, Wallington TJ (2004) Degradation of fluorotelomer alcohols: a likely atmospheric source of perfluorinated carboxylic acids. Environ Sci Technol 38:3316–3321.

Fetzner S (1998). Bacterial dehalogenation. Appl Microbiol Biotechnol 50:633–657.

Gauthier SA, Mabury SA (2005) Aqueous photolysis of 8:2 fluorotelomer alcohol. Environ Toxicol Chem 24:1837–1846.

Gerecke AC, Hartmann PC, Heeb NV, Kohler H-PE, Giger W, Schmid P, Zennegg M, Kohler M (2005) Anaerobic degradation of decabromodiphenyl ether. Environ Sci Technol 39: 1078–1083.

Giesy JP, Kannan K (2001) Global distribution of perfluorooctane sulfonate in wildlife. Environ Sci Technol 35:1339–1342.

Hageman KJ, Istok JD, Field JA, Buscheck TE, Semprini L (2001) *In situ* anaerobic transformation of trichlorofluoroethene in trichloroethene contaminated groundwater. Environ Sci Technol 35:1729–1735.

Hekster F, Laane RWPM, de Voogt P (2003) Environmental and toxicity effects of perfluoroalkylated substances. Rev Environ Contam Toxicol 179:99–121

Hoff PT, Van de Vijver KI, Van Dongen W, Esmans EL, Blust R, De Coen WM (2003) Perfluorooctane sulfonic acid in bib (*Trisopterus luscus*) and plaice (*Pleuronectes platessa*) from the Western Scheldt and the Belgian North Sea: distribution and biochemical effects. Environ Toxicol Chem 22:608–614.

Hoff PT, Scheirs J, Van de Vijver K, Van Dongen W, Esmans EL, Blust R, De Coen W (2004) Biochemical effect evaluation of perfluorooctane sulfonic acid-contaminated wood mice (*Apodemus sylvaticus*). Environ Health Perspect 112:681–686.

Holmes DA, Harrison BK, Dolfing J (1993) Estimation of Gibbs free energies of formation for polychlorinated biphenyls. Environ Sci Technol 27:725–731.

Hopkins GD, McCarty PL (1995) Field evaluation of *in situ* aerobic cometabolism of trichloroethylene and three dichloroethylene isomers using phenol and toluene as the primary substrates. Environ Sci Technol 29:1628–1637.

Huang C-L, Harrison BK, Madura J, Dolfing J (1996) Thermodynamic prediction of dehalogenation pathways for PCDDs. Environ Toxicol Chem 15:824–836.

Hurley MD, Sulbaek Anderson MP, Wallington TJ, Ellis DA, Martin JW, Mabury SA (2004) Atmospheric chemistry of perfluorinated carboxylic acids: reactions with OH radicals and atmospheric lifetimes. J Phys Chem A 108:615–620.

Hyman MR, Page CL, Arp DJ (1994) Oxidation of methyl fluoride and dimethyl ether by ammonia monooxygenase in *Nitrosomonas europaea*. Appl Environ Microbiol 60:3033–3035.

Janssen DB, Dinkla IJT, Poelarends GJ, Terpstra P (2005) Bacterial degradation of xenobiotic compounds: evolution and distribution of novel enzyme activities. Environ Microbiol 7:1868–1882.

Kannan K, Corsolini S, Falandysz J, Fillmann G, Kumar KS, Loganathan BG, Mohd MA, Olivero J, van Wouwe N, Yang JH, Aldous KM (2004) Perfluorooctanesulfonate and related fluorochemicals in human blood from several countries. Environ Sci Technol 38:4489–4495.

Key BD, Howell RD, Criddle CS (1998) Defluorination of organofluorine sulfur compounds by *Pseudomonas* sp. strain D2. Environ Sci Technol 32:2283–2287.

Kissa E (2001) Fluorinated Surfactants and Repellents, 2nd Ed. Dekker, New York.

Koch V, Knaup W, Fiebig S, Geffke T, Schulze D (2007) Biodegradation kinetics of a clariant fluorotelomer-based acrylate polymer: results from a test on aerobic transformation in soil with prolonged exposure. Abstracts, SETAC Europe 17th Annual Meeting, 20–24 May 2007, Porto.

Kondo S, Takahashi A, Tokuhashi K, Sekiya A, Yamada Y, Saito K (2002) Theoretical calculation of heat of formation for a number of moderate sized fluorinated compounds. J Fluor Chem 117:47–53.

Löffler FE, Tiedje JM, Sanford RA (1999) Fraction of electrons consumed in electron acceptor reduction and hydrogen thresholds as indicators of halorespiratory physiology. Appl Environ Microbiol 65:4049–4056.

Marion Meyer JJ, Grobbelaar N, Steyn PL (1990) Fluoroacetate-metabolizing pseudomonad isolated from *Dichaptalum cymosum*. Appl Environ Microbiol 56:2152–2155.

Martin JW, Smithwick MM, Braune BM, Hoekstra PF, Muir DCG, Mabury SA (2004a) Identification of long-chain perfluorinated acids in biota from the Canadian Arctic. Environ Sci Technol 38:373–380.

Martin JW, Kannan K, Berger U, de Voogt P, Field JA, Franklin J, Giesy JP, Harner T, Muir DC, Scott B, Kaiser MA, Järnberg U, Jones KC, Mabury SA, Schröder HF, Simcik M, Sottani C, van Bavel B, Kärrman AH, Lindström G, van Leeuwen SP (2004b) Analytical challenges hamper perfluoroalkyl research. Environ Sci Technol 38:248A–255A.

Masunaga S, Susarla S, Yonezawa Y (1996) Dechlorination of chlorobenzenes in anaerobic estuarine sediment. Water Sci Technol 33:173–180.

McCarty PL (1997) Microbiology: breathing with chlorinated solvents. Science 276:1521–1522.

Meesters RJW, Schröder HFr (2004) Perfluorooctane sulfonate: a quite mobile anionic anthropogenic surfactant, ubiquitously found in the environment. Water Sci Technol 50:235–242.

Natarajan R, Azerad R, Badet B, Copin E (2005) Microbial cleavage of C-F bond. J Fluorine Chem 126:425–436.
Olsen GW, Huang HY, Helzlsouer KJ, Hansen KJ, Butenhoff JL, Mandel JH (2005) Historical comparison of perfluorooctanesulfonate, perfluorooctanoate, and other fluorochemicals in human blood. Environ Health Perspect 113:539–545.
Oltmanns RH, Müller R, Otto MK, Lingens F (1989) Evidence for a new pathway in the bacterial degradation of 4-fluorobenzoate. Appl Environ Microbiol 55:2499–2504.
Prevedouros K, Cousins IT, Buck RC, Korzeniowski SH (2006) Sources, fate and transport of perfluorocarboxylates. Environ Sci Technol 40:32–44.
Remde A, Debus R (1996) Biodegradability of fluorinated surfactants under aerobic and anaerobic conditions. Chemosphere 32:1563–1574.
Renner R (2001) Growing concern over perfluorinated chemicals. Environ Sci Technol 35:154A–160A.
Rhoads KR, Janssen EM-L, Luthy RG, Criddle CS (2008) Aerobic biotransformation and fate of N-ethyl perfluorooctane sulfonamidoethanol (N-EtFOSE) in activated sludge. Environ Sci Technol 42:2873–2878.
Sáez M, de Voogt P, Parsons JR (2008) Persistence of perfluoroalkylated substances in closed bottle tests with municipal sewage sludge. Environ Sci Pollut Res (in press).
Schröder HF (2003) Determination of fluorinated surfactants and their metabolites in sewage sludge samples by liquid chromatography with mass spectrometry and tandem mass spectrometry after pressurised liquid extraction and separation on fluorine-modified reversed-phase sorbents. J Chromatogr A 1020:131–151.
Schultz MM, Barofsky DF, Field JA (2003) Fluorinated alkyl surfactants. Environ Eng Sci 20:487–501.
Semprini L (1997) Strategies for the aerobic co-metabolism of chlorinated solvents. Curr Opin Biotechnol 8:296–308.
Skoczynska EM, Zegers B, de Voogt P, Parsons JR (2005) Reductive debromination of polybrominated diphenyl ethers (PBDEs) by anaerobic sediment microorganisms. Organohalogen Compd 67:572–574.
Smidt H, de Vos WM (2004) Anaerobic microbial dehalogenation. Annu Rev Microbiol 58:43–73.
Sonier DN, Duran DL, Smith GB (1994) Dechlorination of trichlorofluoromethane (CFC-11) by sulfate-reducing bacteria from an aquifer contaminated with halogenated aliphatic compounds. Appl Environ Microbiol 60:4567–4572.
Speight JG (2005) Lange's Handbook of Chemistry, 16th Ed. McGraw-Hill, New York.
Suflita JM, Horowitz A, Shelton DR, Tiedje JM (1982) Dehalogenation: a novel pathway for the biodegradation of haloaromatic compounds. Science 218:115–117.
van Eekert MHA, Stams AJM, Field JA, Schraa G (1999) Gratuitous dechlorination of chloroethanes by methanogenic granular sludge. Appl Microbiol Biotechnol 51:46–52.
Vargas C, Song B, Camps M, Haggblom MM (2000) Anaerobic degradation of fluorinated aromatic compounds. Appl Microbiol Biotechnol 53:342–347.
Wang N, Szostek B, Folsom PW, Sulecki LM, Capka V, Buck RC, Berti WR, Gannon JT (2005a) Aerobic biotransformation of $^{14}$C-labeled 8-2 telomer B alcohol by activated sludge from a domestic sewage treatment plant. Environ Sci Technol 39:531–538.
Wang N, Bogdan Szostek B, Buck RC, Folsom PW, Sulecki LM, Capka V, Berti WR, Gannon JT (2005b) Fluorotelomer alcohol biodegradation: direct evidence that perfluorinated carbon chains breakdown. Environ Sci Technol 39:7516–7528.
Yamada T, Lay TH, Bozzelli JW (1998) Ab initio calculations and internal rotor: contribution for thermodynamic properties $S°_{298}$ and $C_p(T)$'s (300 < $T$/K < 1500): group additivity for fluoroethanes. J Phys Chem A 102:7286–7293.
Yamada T, Bozzelli JW, Berry RJ (1999) Thermodynamic properties ($\Delta H_{f(298)}$, $S_{(298)}$, and $C_p(T)$ (300 ≤ $T$ ≤ 1500)) of fluorinated propanes. J Phys Chem A 103:5602–5610.
Yan D-Z, Liu H, Zhou N-Y (2006) Conversion of *Sphingobium chlorophenolicum* ATCC 39723 to a hexachlorobenzene degrader by metabolic engineering. Appl Environ Microbiol 72:2283–2286.

# Lead Stress Effects on Physiobiochemical Activities of Higher Plants

Rakesh Singh Sengar, Madhu Gautam, Rajesh Singh Sengar, Sanjay Kumar Garg, Kalpana Sengar, and Reshu Chaudhary

## Contents

| | | |
|---|---|---|
| 1 | Introduction | 73 |
| 2 | Sources of Lead Pollution | 75 |
| 3 | Behavior of Lead in Plants and Effects on Plant Processes | 76 |
| | 3.1 Foliar Deposition of Lead | 76 |
| | 3.2 Lead Uptake by Plants | 76 |
| | 3.3 Localization and Levels of Lead in Plant Parts | 77 |
| | 3.4 Seed Germination and Seedling Growth | 80 |
| | 3.5 Plant Growth and Productivity | 80 |
| | 3.6 Effects on Photosynthesis | 81 |
| | 3.7 Effects on Respiration | 84 |
| | 3.8 Effects on Nitrogen Assimilation | 85 |
| | 3.9 Pigment Control | 86 |
| 4 | Summary | 86 |
| | References | 87 |

## 1 Introduction

Environmental degradation caused by rapid industrial expansion of activities such as mining, power generation, transportation, and intensive agriculture, among others, has become a major threat to the sustenance and welfare of mankind. During the past

---

M. Gautam
Institute of Management Studies, C-238, Bulandshahr Road, Indl, Area, Lal Quan, G.T. Road, Ghaziabad, India

R.S. Sengar, K. Sengar, R. Chaudhary
College of Biotechnology, Sardar Vallabh Bhai Patel University of Agriculture and Technology, Meerut, India

R.J. Sengar, R.S. Sengar
Meerut Institute of Engineering and Technology, Partapur Bypass, Meerut, India

S.K. Garg
Department of Plant Science, M.J.P. Rohillkhand University Bareilly, India

few decades, heavy metals, which constitute important groups of environmental pollutants, have received increasing attention. Heavy metals are nondegradable and readily enter the food chain and can subsequently endanger human and animal health (Valle and Ulmer 1972; Niragh 1988; Schmidt and Ibrahim 1994; Liou et al. 1974). Contamination of the environment by heavy metals further jeopardizes human welfare by reducing productivity of agricultural land and curtailing crop yields (Strand et al. 1990). Lead (Pb) is one of the most abundant heavy metals and is highly phytotoxic. Although lead is not an essential element for plants or animals, both often absorb lead easily. Uptake of lead and its accumulation in plants occur either directly, when it is absorbed via roots along with minerals and water, or when absorbed from air through plant shoots and foliage. Lead contamination of soil, water, and air has increased rapidly because of industrialization and urbanization during the past few decades. The general trend is that heavy metal enrichment is occurring in urban > rural > remote locations. Lead contamination of air, water, and soil is also occurring from industrial point sources. Such pollution correlates well with vehicular traffic volume, presence of metal working industries, and distance from metal generating and processing sites (Daines et al. 1970; Smith 1972; Wheeler and Rolfe 1979; Pilgrim and Hughes 1994).

Lead accumulates primarily in the surface soil layer, and its concentration decreases with soil depth (de Abreu et al. 1998). Lead may exist in the atmosphere as dust, fumes, mists, and vapors and in soil as a mineral. Soils along roadsides are particularly lead rich because of pollution from vehicles burning leaded gasoline. Other important sources of lead pollution are geological weathering, industrial processing of ores and minerals, leaching of metals from solid wastes, and waste from animal and human excreta. In moist soil, except near lead ore deposits, lead concentrations range between 2 and 200 ppm (Wright et al. 1995), with an average of about 16 ppm (De Treville 1964). Most lead is tightly and irreversibly bound to ion-exchangeable soil surfaces. However, plant roots are apparently able to extract some lead from soil, which may then accumulate in plant tissues. In pinto bean roots, for example, lead is accumulated up to 32.4 mg/g dry weight. Because one cannot easily eliminate lead from modern human life, soil contamination with lead is not likely to decrease in the near future (Yang et al. 2000).

Toxic effects of lead pollution are well established. The main sources of lead intake by animals are food, lead-contaminated dusts, and air. Lead affects the animal nervous system, reproductive system, cardiovascular system, and hemoglobin biosynthesis (Jawarski 1978; Liou et al. 1974). The effects of soil and waterborne lead pollution on plants have been widely studied, but such studies have largely been confined to identifying morphological and physiological changes in plants (Foy et al. 1978; Koeppe 1981). Soils contaminated with lead cause sharp decreases in crop productivity, thereby posing a serious problem for agriculture (Johnson and Eaton 1980).

In this review, we focus on the sources of lead, its uptake and transport within plants, and the physiological, biochemical, and ultrastructural changes in plants caused by lead toxicity. Moreover, it is an objective of this review to examine and present an overview of the physiological and biochemical changes in plant metabolic activities caused by lead.

## 2 Sources of Lead Pollution

The major sources of lead environmental contamination are metal smelters, use of lead arsenate pesticides, and phosphate fertilizers and $Pb^{2+}$-based paints (Goldsmith et al. 1976; Eick et al. 1999). Residues or particles of lead from lead-based paints may be hazardous, and incidents of such poisoning are widely reported in both underdeveloped and developed societies (Lagerwerff et al. 1973). Metallic lead from leaded gasoline is emitted from automobile exhaust as lead aerosols (Smith 1972; Eick et al. 1999). Such combustion of lead alkyl derivatives in gasoline causes extensive contamination along roadsides (Canon and Bowler 1962; Chow 1970; Eick et al. 1999; Pages and Ganje 1970). Plants growing along roadsides are therefore enriched with lead and may directly or indirectly contribute to general atmospheric lead contamination. Lead occurs in the environment mainly as acidic salts. Figure 1 depicts various sources that contribute to lead pollution in the environment.

Among the most prominent sources of lead release to the environment are automobile emissions (Canon and Bowler 1962), lead-based paints, fertilizing with sewage sludge, or watering with sewage water (Hinsely et al. 1971; Chaney and Ryan 1994; Paivoke 2002), particulate metal oxides from vehicles, industrial processing of ores and minerals (Purves 1977), and smelting of lead. Lead smelters

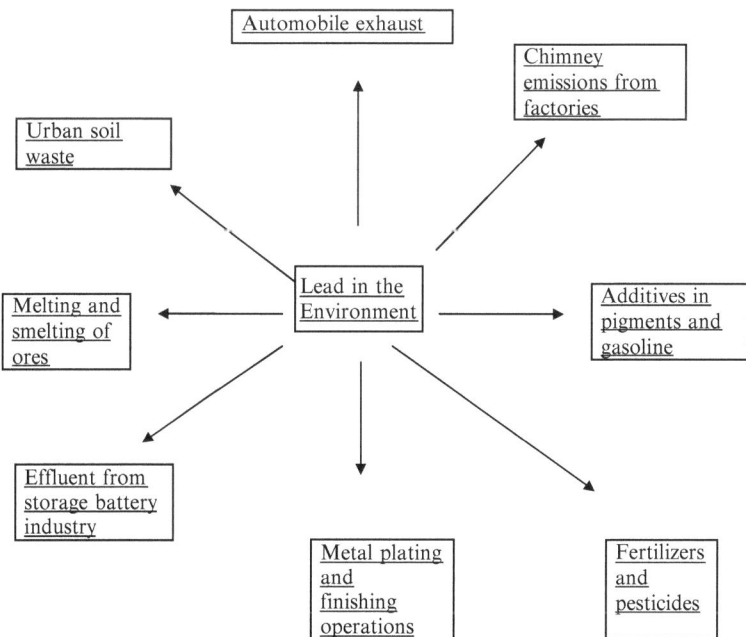

**Fig. 1** Sources of lead pollution in the environment

may release lead for deposition in amounts ranging between 0.046 and 20.1 kg/ha/yr (Pilgrim and Hughes 1994). Mine water also transports large amounts of fine particulate sediments contaminated with lead (Laxen and Harrison 1977). Other sources of contamination result from use of lead arsenate insecticides, phosphate fertilizers, and spent lead shot from hunting activities (Eick et al. 1999; Goldsmith et al. 1976). Finally, rock phosphate, used in fertilizers, contains heavy metals including lead as impurities.

## 3 Behavior of Lead in Plants and Effects on Plant Processes

### 3.1 Foliar Deposition of Lead

Lead deposited on plants exists largely as superficial deposits or a topical aerosol coating on plant surfaces (Zimdahl 1976). Lead deposited on foliage as a topical coating may comprise 5–200 times the amount remaining after such crop foliage surfaces are washed (Smith 1972). Pubescent leaves retain more lead than do smooth leaf surfaces, although such residues may be washed off by rainwater (Laxen and Harrison 1977; Zimdahl and Koeppe 1977). Two species of plants, *Cassia tora* and *Cassia occidentalis*, that commonly grow along roadsides are known to accumulate up to 300 ppm lead (Krishnayapa and Bedi 1986).

### 3.2 Lead Uptake by Plants

Lead residues in soil exist as soluble and insoluble salts. Soils often contain lead in the range of 400–800 mg kg$^{-1}$, whereas in industrial areas lead levels in soil may reach 1,000 mg kg$^{-1}$ (Angelone and Bini 1992). Plants may absorb lead either from air or from soil. Downy leaves absorb heavy metals from the atmosphere (Godzik 1993). Lead may be absorbed by plants and taken up into leaves at high levels from low-fertility soil, particularly by susceptible plant species (Johnson and Proctor 1977; Johnson et al. 1977). Airborne lead exists mainly as a component of dust deposited on plant leaves or other exposed surfaces (Zimdahl and Koeppe 1977). Several factors affect the amount and speed of uptake and transpiration of lead into plants. Among these are soil particle size, soil cation-exchange capacity, root surface area, root exudates that solubilize iron, manganese, lead, etc., and the presence of mycorrhizae (Davies 1995). Plants growing along roadsides are exposed to high lead levels because of their proximity to vehicles that combust leaded fuel. As expected, lead levels in such plants tend to decrease as their distance from a highway increases (Wallace et al. 1976; Goldsmith et al. 1976). The lead concentration in grass collected along a 10–70 m section of railway ranged between 23 and 228 ppm (Pilgrim and Hughes 1994).

Most soils (not located near lead-containing ore deposits) have lead concentrations that range between 2 and 200 ppm (Wright et al. 1995), with an average of about 16 ppm (De Treville 1964). Lead is generally tightly bound to colloidal organic molecules in soil. Plant roots are able to extract some of this bound lead, which may then accumulate in plant tissue. However, the bulk of plant-absorbed lead remains in roots (Kumar et al. 1995).

As mentioned, various soil conditions affect soil uptake of lead; such conditions include soil cation-exchange capacity (Baumhardt and Welch 1972; Miller et al. 1973), organic matter content (Liebhardt and Koske 1974), and soil pH (Cox and Rains 1972; Lagerwerff et al. 1973; Miller et al. 1975; Zimdahl and Faster 1976; John and Van Laerhovexn 1972; Karawmanos et al. 1976). The adsorption of elemental lead by soil follows the Langmuir relationship, which portends increased uptake with increasing pH between 3.0 and 8.5 (Lee et al. 1998). However, Blaylock et al. (1997) reported that, in soil with pH from 5.5 to 7.5, lead solubility is controlled by phosphate or carbonate precipitates that constrain plant lead uptake even if plants have the capacity to accumulate it.

Lead is more toxic to corn, beans, *Lactuca* sp., and radishes when these plants are grown on calcareous soils (Bala and Setia 1990). Addition to soil of phosphate lime, organic matter, and chloride is known to reduce plant lead uptake (Liebhardt and Koske 1974). Lead uptake by corn and sunflower is directly related to plant transpiration rates and reduction of photosynthesis (Bazzaz et al. 1975). Lead absorption may also be affected by the life stage of plants and passive environmental factors. Young tobacco seedlings accumulated lead more rapidly than did older plants (Zimdahl and Koeppe 1977). An abundance of lead pyrophosphate and lead orthophosphate in soil reduces lead absorption. Uptake of heavy metals from soil is also modified by presence of chelating agents (Wallace et al. 1976). For example, ethylenediaminetetraacetic acid (EDTA) increases lead uptake (Martin and Hammond 1966) in barley plants. Diethylenetriaminepentaacetic acid (DTPA) increases the uptake and translocation of lead (Patel et al. 1977). In general, dicots accumulate significantly higher amounts of lead in the roots than do monocots. Marschner et al. (1996) demonstrated that ectomycorrhizae influenced uptake, transport, and toxicity of lead in Norway spruce plants.

## 3.3 *Localization and Levels of Lead in Plant Parts*

Transport within and cellular localization of lead in plants is important in its toxic manifestation. For example, lead residues stored in vacuoles may not cause deleterious effects on the plant. Lead absorbed by roots through root hairs reaches the vascular system via the cortex and endodermis. The addition of synthetic chelates, such as *N*-(4-hydroxyethyl) ethylenediaminetriacetic acid (H-EDTA) or EDTA, in combination with low pH, effectively prevents cell wall retention of lead, making it available for translocation to shoots (Jarvis and Leung 2002). Lead accumulates and localizes in the cell wall of the xylem (Glater and Hernandez 1972; Lane and

Martin 1977; Sieghardt 1984). In general, dicots accumulate significantly higher amounts of lead in roots than do monocots (Huang and Cunningham 1996). Lead deposited in other parts of vascular tissue are often further transported to shoots and other aboveground plant structures (Jones et al. 1973; Hardimann et al. 1984).

A voltage-gated calcium (Ca) channel in the root cell plasma membrane has been characterized in wheat and corn plants (Marshall et al. 1994; Huang et al. 1994). Huang and Cunningham (1996) found that lead inhibited voltage-gated Ca channel activity in the plasma membrane of wheat roots. Such inhibition could arise from direct lead blockage of the channel or may be from competitive inhibition of transport through the Ca channel. While monitoring lead transport into isolated cells, Tomsig and Suszkiw (1991) observed permeation of lead through Ca channels. These authors also found that voltage-gated lead transport was blocked by nifedipine (a Ca-channel blocker) and enhanced by BAY K86 44 (a Ca-channel agonist).

Lead is absorbed through roots and slowly accumulates in cells; absorbed lead may migrate to cell walls outside the plasma lemma (Malone et al. 1974). Miller et al. (1973) demonstrated that *Zea mays* L. plants could translocate and accumulate significant quantities of lead in leaves in a concentration-dependent manner. Lead also accumulates in plant intercellular spaces, endoplasmic reticulum, dictyosome and dictyosome vesicles, and even in cell nuclei (Tandler and Solavi 1969). In onion root tips, the metal localizes as Pb orthophosphate in the nucleus (Tandler and Solavi 1969). In onion (*Allium cepa*) plants, absorbed lead is localized at its highest concentration in root tips, followed by proximal parts of the root, while the lowest concentration is found in the root base (Wierzbicka 1998). Godzik (1993) found that the highest lead content exists in senescing leaves and the least in young leaves (Godzik 1993). The cell wall and vacuole of onion (*Allium cepa*) together account for about 96% of absorbed lead (Wierzbicka 1994). Lead binds strongly to carboxyl groups of the carbohydrates galacturonic and glucuronic acids in the cell wall, which restricts its apoplastic transport (Rudakova et al. 1988).

Lead accumulates to high levels in organs of susceptible species (Johnson and Procter 1977; Johnson et al. 1977). The limited transport of lead from roots to other plant organs is the result of the barrier that exists in the root endodermis. It appears that casparian strips of the endodermis are the major factor restricting lead transport across the endodermis into central cylinder tissue (Seregin et al. 2004). According to Lane and Martin (1977), the endodermis appears to act only as a partial barrier, as some lead moves up the plant in vascular structures and diffuses out into surrounding tissues. Such transport in plants provides evidence that some lead moves via the symplast. It is concluded, however, that lead in roots moves primarily via the apoplast; this conclusion is supported by the report that most residual lead is readily extractable in water (Broyer et al. 1972). The possibility of symplastic transport of lead has been demonstrated in onion roots and garden cress hypocotyls (Wierzbicka 1987).

Concentration-dependent lead accumulation has been reported in maize and pea leaves (Sinha et al. 1988a,b) and in sesame roots and leaves (Kumar et al. 1992). More lead accumulated in roots and leaves of *Sesamum indicum* H7-1, a sensitive variety (Kumar et al. 1992) than in *Sesamum indicum* Cv. PB-1, a tolerant sesame cultivar. Seregin et al. (2004) demonstrated that a significant amount of lead was retained at the surface of the plasma lemma rather than in cell walls. Lead enters injured cells accompanied by compounds such as procion dyes, which are not absorbed by undamaged cells (Seregin et al. 2004). The pattern of lead distribution in roots differs depending on whether concentrations are lethal (Seregin et al. 2004). Suchodoller (1967) found that lead applied to barley was retained in root epidermis while only a small amount was found in vascular tissues, which suggests that the extent of localization of lead in different tissues is plant species dependent. Once the testa is ruptured, lead is taken up very rapidly, with notable exceptions in meristematic regions of the radicle and hypocotyl (Lane and Martin 1977). In cotyledons, lead moves through vascular tissues and accumulates in discrete distal parts (Lane and Martin 1977). In leaf cells of *Potamogeton* spp., it was shown that the electrochemical gradients between cell vacuoles and their bathing solution ranged from 150 to 240 mV (Denny and Weeks 1968), which may favor a passive influx of lead into vacuoles during lead exposure. In *Stigeoclonium*, lead concentration in such vacuoles may protect the cell contents from lead toxicity (Silverberg 1975). *Lemna miner* L. (duckweed) plants, when treated with lead for 1 hr, showed the maximum concentration of lead in small vacuoles (Samardakiewicz and Wozny 2000). The localization of lead in vacuoles and cell walls may foster increased apoplastic transport and result in redistribution of lead within the plant (Samardakiewicz and Wozny 2000). A generalized view of the effects of lead toxicity on key physiological processes in plants is presented in Fig. 2.

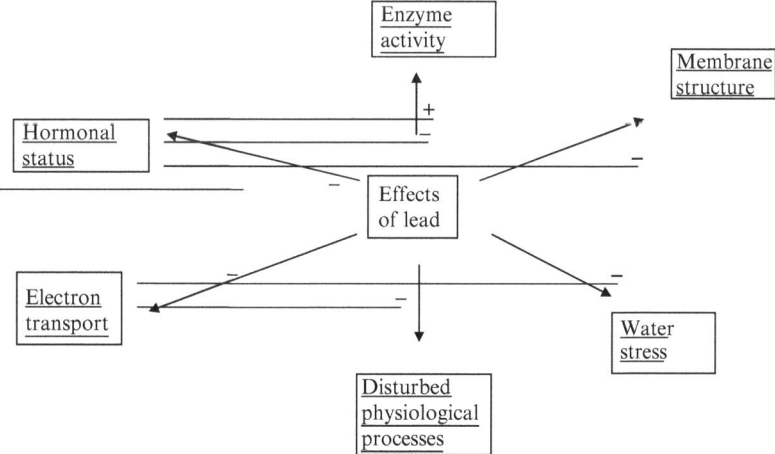

**Fig. 2** The diagram is a generalized view of lead toxicity in plants. Lead phytotoxicity causes dehydration and water stress (see Fig. 3), alteration in membrane permeability, decreases in hormonal status and electron transport activities, and either enhances or inhibits enzymes. These events ultimately result in disturbed physiological processes (+ and − signs indicate positive and negative effects, respectively)

## 3.4 Seed Germination and Seedling Growth

Seed germination is the initial event in the life of plants. Lead is known to inhibit the seed germination of *Spartiana alterniflora* (Morzek and Funicelli 1982), *Pinus helipensis* (Nakos 1979), rice (Mukherji and Maitra 1976; Wozny et al. 1982), and *Zea mays* (Iqbal and Mushtaq 1987). Inhibition of germination may result from the interference of lead with important enzymes. Mukherji and Maitra (1976) observed that 60 mM lead acetate inhibited protease and amylase by about 50% in rice endosperm.

Early seedling growth was also inhibited by lead in soybean (Huang et al. 1974), rice (Mukherji and Maitra 1976), maize (Miller et al. 1975), barley (Stiborova et al. 1987), tomato, eggplant (Khan and Khan 1983), and certain legumes (Sudhakar et al. 1992). Lead also inhibited root and stem elongation and leaf expansion in *Allium* species (Gruenhale and Jager 1985), barley (Juwarkar and Shende 1986), and *Raphanus sativas*. The degree to which root elongation is inhibited depends upon the concentration of lead and ionic composition and pH of the medium (Goldbold and Hutterman 1986). Concentration-dependent inhibition of root growth has been observed in *Agrostis capillaries* seedlings (Symenoidis et al. 1985) and in *Sesamum indicum* (L) Cv. HT-1 (Kumar et al. 1992).

## 3.5 Plant Growth and Productivity

A high lead level in soil induces abnormal morphology in many plant species. For example, lead causes irregular radial thickening in pea roots, cell walls of the endodermis, and lignification of cortical parenchyma (Paivoke 1983). Lead also induces proliferation effects on the repair process of vascular plants (Kaji et al. 1995). Lead administered to potted sugarbeet plants at rates of 100–200 ppm caused chlorosis and growth reduction (Hewitt 1953). In contrast, there was no visual symptom of lead toxicity in alfalfa plants exposed to 100 mg/mL (Porter and Cheridan 1981). In another alfalfa study, lead sulfate or lead nitrate at 250 ppm in soil had no effect on alfalfa yield (Taylor and Allinson 1981). The absence of effects on alfalfa was attributed to poor absorption of lead by root cells (Lagerwerff et al. 1973) and poor translocation of lead within the plants (Karawmanos et al. 1976). Foliar application of lead was observed to reduce growth and yield in wheat (Rashid and Mukherji 1993).

Low amounts of lead (0.005 ppm) caused significant reduction in growth of lettuce and carrot roots (Baker 1972). The fresh weight of barley was increased 25% and 67% in the presence of 0.3% and 3.0% lead, respectively, when used as a polymicrofertilizer (Keaton 1937). In soybean (*Glycine max*), 300 mM $Pb^{2+}$ inhibited pod fresh weight by 35.1% (Huang et al. 1974), and decreased the yield of tomatoes by 25% (Khan and Khan 1983).

Observations show that lead exposure inhibited root growth and ultimately caused plant death in several species (Bradshaw 1952; Bradshaw and McMeilly

1981; Symeonidis et al. 1985). It has been suggested that lead affects plant growth by constraining availability of essential elements in *Allium ponum* and *Pisum sativum* Gruenhage. Inhibitory effects of $Pb^{2+}$ on growth and biomass production may possibly derive from effects on metabolic plant processes (Van Assche and Clijesters 1990). Some researchers believe the primary cause of cell growth inhibition arises from a lead-induced stimulation of indole-3-acetic acid (IAA) oxidation (Mukherji and Maitra 1976; Burzynski and Jakob 1983). The inhibitory effect of lead on growth may also arise from lead interference with auxin-regulated cell elongation, which can be demonstrated in an *Avena* coleoptile assay (Lane et al. 1978). Lead is also known to decrease the growth of cauliflower, spinach, and parsley (Salim et al. 1995).

## 3.6 Effects on Photosynthesis

Plant photosynthesis is inhibited by higher concentrations of lead; examples include soybean (Bazzaz et al. 1974), spruce (Keller and Zuber 1970), sunflower (Bazzaz et al. 1975), and American sycamore (Carlson and Bazzaz 1977). Reduction in photosynthesis may result from effects on stomata (Bazzaz et al. 1974; Rolfe and Bazzaz 1975), disruption of the chloroplastic organization (Rebechini and Hanzely 1974; Kumar 1990), or from biochemical changes in the metabolic products of photosynthesis (Hampp et al. 1973; Bazzaz and Govindji 1974; Sarkar and Jana 1987; Jana et al. 1987). Lead is also known to affect photosynthesis by inhibiting activity of carboxylating enzymes (Stiborova et al. 1987). Lead inhibits photosynthesis up to approximately 60% (Miles et al. 1972).

Lead inhibits chlorophyll synthesis by causing impaired uptake of essential elements such as Mg and Fe (Burzynski 1987). Lead also damages the photosynthetic apparatus by its affinity for protein N- and S-ligands (Ahmed and Tajmir-Riahi 1993). Inhibition of total chlorophyll by lead was observed in oats (Fiussello and Molinari 1973), aquatic plants (Jana and Choudhari 1984), cucumber (Burzynski 1985), giant dodder (*Cascuta reflexa*) (Jana et al. 1987), mungbean seedlings (Prassad and Prassad 1987), and maize and pea (Sinha et al. 1988a; Sinha et al. 1988b). Enzymes involved in synthesis of chlorophyll are also affected by lead (Burzynski 1985; Prassad and Prassad 1987; Sinha et al. 1988a,b) in plants as well as in algae (De Filippis and Pallaghy 1976; De Filippis et al. 1981). An enhancement of chlorophyll degradation occurs in lead-treated plants because of increased chlorophyllase activity (Drazkiewicz 1994). Chlorophyll *b* is reported to be more susceptible than chlorophyll *a* to lead treatment (Vodnik et al. 1999).

Lead exposure is also known to alter the lipid composition of thylakoid membranes (Stefanov et al. 1995). According to Kosobrukhov et al. (2004), the photosynthetic activity of plants is governed by many factors including stomatal cell size, number of stomata, stomatal conductance, and leaf area. While studying the effects of lead on the development of thylakoid of cucumber and poplar plants,

Sarvari et al. (2002) observed increased chlorophyll content either in the photo system (PS) II core or light-harvesting chlorophyll (LHC) II at low lead concentrations, whereas a strong decrease in seedling chlorophyll level was seen at a lead concentration of 50 mM. At this level, amounts of lead inside the leaf may have directly caused inhibition of chlorophyll synthesis (Sengar and Pandey 1996). A strong relationship exists between amounts of lead applied and degree of whole plant photosynthetic inhibition, which is believed to result from stomatal closure rather than direct lead effects on photosynthesis (Bazzaz et al. 1975). Effects of lead on various components of photosynthesis, mitotic irregularities, respiration, water regime, and nutrient uptake are shown in Fig. 3.

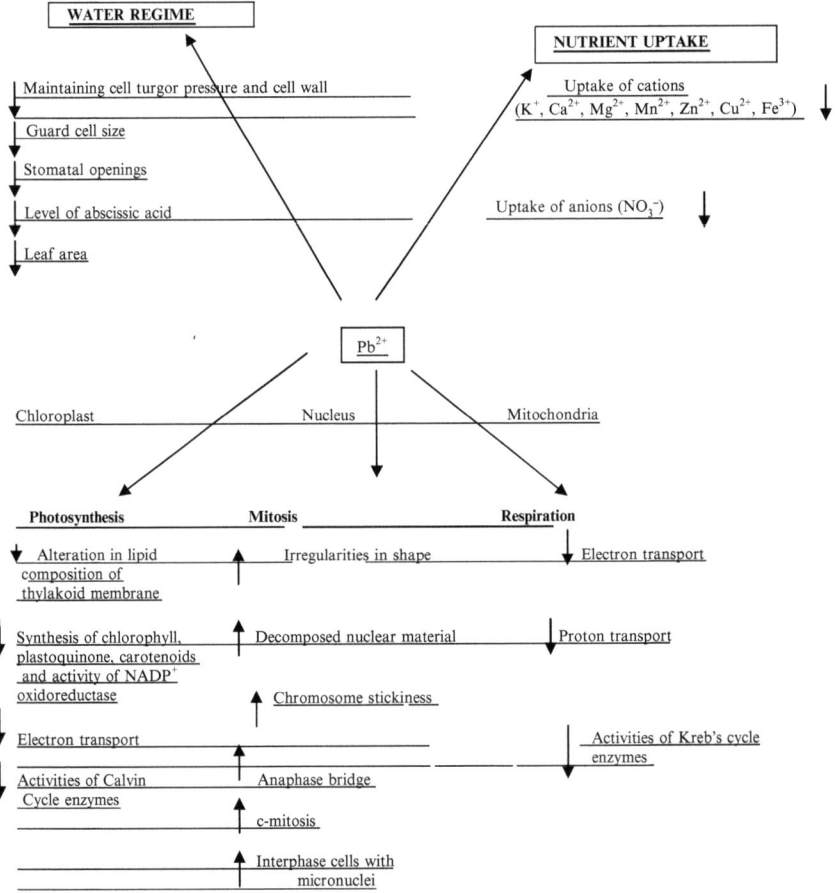

**Fig. 3** Effects of lead on photosynthesis, mitosis, respiration, water regime, and nutrient uptake (*arrows* represent enhanced and/or decreased activities, respectively)

Lead also inhibits electron transport (Rashid et al. 1994). The inhibition of the Hill reaction has been reported in the angiosperm parasite *Cuscuta reflexa* (Jana et al. 1987). Bazzaz and Govindji (1974) observed that lead chloride either stimulated or inhibited PS II activity, depending on the pH of the reaction medium. Ribulose-1,5-bisphosphate carboxylase (RUBP) and ribulose-1,5-bisphosphate kinase (Table 1), the two enzymes involved in photosynthetic $CO_2$ assimilation, are

**Table 1** Effect of lead on the activity of enzymes in higher plants

| Enzyme | Plant | Lead concentration | Effect | Reference |
|---|---|---|---|---|
| Acid phosphatase | Soybean leaves | 100 ppm | Increase | Lee et al. (1976) |
| | Maize roots | 200 & 500 ppm | Decrease | Maier (1978) |
| | Maize leaves | 200 & 500 ppm | Increase | Maier (1978) |
| | *Zea mays* seedling | 100 ppm | Increase | Lee et al. (1976) |
| | | 5–100 ppm | Decrease | Iqbal and Mushtaq (1987) |
| Esterases | *In vitro* extract from maize | 200–500 ppm | Decrease | Maier (1977) |
| | Alfalfa | 200–500 ppm | | |
| | *Zebrina pendla* | 200–500 ppm | | |
| Glutamine synthetase | Soybean leaves | 100 ppm | Decrease | Lee et al. (1976) |
| Malic dehydrogenase | Soybean leaves | 100 ppm | Increase | Lee et al. (1976) |
| | Alfalfa roots | 500 ppm | Decrease | Maier (1977) |
| | Alfalfa roots | 500 ppm | Increase | Maier (1977) |
| | *Zebrina pendula* | 50–500 ppm | Increase | Maier (1977) |
| Nitrogenase | *Zostera marina* roots | 10 ppm | No effect | Brackup and Caponc (1985) |
| | | 100 ppm | No effect | Brackup and Capone (1985) |
| Peroxidases | Soybean leaves | 100 ppm | Decrease | Lee et al. (1976) |
| Pyruvate kinase | Spinach extracts | 100 μM | Increase | Hampp et al. (1973) |
| Rubolose 1,5-bisphosphate carboxylase | Spinach extracts | 5 μM | Increase | Hampp et al. (1973) |
| | Maize seedlings | | | |
| Nitrate reductase | Sorghum | 10–100 μM | Decrease | Venketraman et al. (1978) |
| | Maize leaves | 0.1 mM | Increase or Decrease | Sinha et al. (1988b) |
| | | 10 mM | | Sinha et al. (1988b) |
| Ribulose-5-phosphate kinase | Spinach extracts | 5 μM | Decrease | Hampp et al. (1973) |

inhibited by 5 mM $Pb^{2+}$ in spinach extract (Hampp et al. 1973), although the RUBP carboxylase and phosphenol/pyruvate carboxylase activities are known to be inhibited by lead (Stiborova et al. 1987). Lead effects have been reported for both donor and acceptor sites of PS II, the cytochrome *b/f* complex, and PS I. It is largely accepted that PS I electron transport is less sensitive to inhibition by lead than is PS II (Mohanty et al. 1989; Seresen et al. 1998). Lead also causes strong dissociation of the oxygen-evolving extrinsic polypeptide of PS II and displacement of Ca, Cl, and Mn from the oxygen-evolving complex (Rashid and Mukerji 1991). Ahmed and Tajmir-Riahi (1993) found conformational changes in LHC II subunits, following binding with lead *in vitro*.

## 3.7 Effects on Respiration

Lead has been found to increase dark respiration in soybean (Lee et al. 1976). Glyceraldehyde-3-$PO_4$ dehdrogenase from spinach, and malate dehdrogenase from soybean and alfalfa leaves, showed activity increases at low $Pb^{2+}$ concentrations (Hampp et al. 1973; Lee et al. 1976; Maier 1977). *In vitro* application of lead to plant cell mitochondrial preparations revealed a decrease in respiration rate with increasing lead concentrations (Reese and Roberts 1985). Using isolated chloroplasts and mitochondria from different plant species, it has been shown that lead affects the flow of electrons in the electron transport system (Miles et al. 1972; Bazzaz and Govindjee 1974).

Plant respiration is also affected by lead. For example, a 3-hr treatment of maize root tips with 20 mM lead inhibits respiration by 28% –40% (Koeppe 1977). However, respiration in soybean leaves increased by 22% at a lead concentration of 100 mg/ml (Lee et al. 1976). At lower lead concentrations, stimulation of respiration is observed in whole plants (Lee et al. 1976), detached leaves (Lamoreaux and Chaney 1978), isolated protoplasts (Parys et al. 1998), and mitochondria (Koeppe and Miller 1970). When detached leaves from $C_3$ plants (pea, barley), and $C_4$ plants (maize) were exposed to 5 mM lead nitrate for 24 hr, the respiratory rate was stimulated by 20% –50% (Romanowska et al. 2002). The inhibitory effect of lead at higher concentrations appears result from uncoupling of oxidative phosphorylation (Miller et al. 1973). Table 2 presents an overview of the effects of lead on activities of various enzymes in different plant species.

Lead has a very high passive affinity for isolated mitochondria and causes drastic structural damage in algae at 5 mM. At 100 mM, lead caused deformation and structural alterations of nuclei, structural alterations of mitochondria, and increased autocatalytic activity (Roderer 1984). There is evidence that lead binds with membranes and that the binding causes physiological effects. Mitochondria isolated from lead-treated pea leaves oxidized substrates (glycine, succinate, malate) at higher rates than mitochondria from control leaves (Romanowska et al. 2002).

**Table 2** Effect of lead on enzyme activity of different metabolic processes

| Metabolic processes | Enzymes | Plant species | Effect of lead | Reference |
|---|---|---|---|---|
| Chlorophyll synthesis | δ-Aminolevulinate | *Pennisetum typhoideum* | – | Prassad and Prassad (1987) |
| $CO_2$ fixation | Ribulose-1,5-bis-phosphate | *Avena sativa* | – | Hampp et al. (1973) |
| $N_2$ assimilation | Nitrate reductase | *Cumcumis sativus* | – | Sinha et al. (1988b) |
| | Glutamate synthetase | *Gycine max* | – | Lee et al. (1976) |
| Protein hydrolysis | Protease | *Hydrilla verticillata* | + | Jana and Choudhary (1984) |
| Phosphohydrolase | Alkaline phosphatase | *Hydrilla verticillata* | + | Jana and Choudhary (1984) |
| | Acid phosphatase | *Gycine max* | + | Lee et al. (1976) |
| Sugar metabolism | α-Amylase | *Oryza sativa* | – | Mukherji and Maitra (1976) |
| Antioxidative metabolism | Catalase | *Oryza sativa* | – | Verma and Dubey (2003) |
| | Guaiacol peroxidase | *Gycine max* | + | Lee et al. (1976) |
| | Ascorbate oxidase | *Phaseolus aureus* | + | Rashid and Mukherji (1991) |
| | Ascorbate peroxidase | *Oryza sativa* | + | Verma and Dubey (2003) |
| | Glutathione reductase | *Oryza sativa* | + | Verma and Dubey (2003) |
| | Superoxide dismutase | *Oryza sativa* | + | Verma and Dubey (2003) |

## 3.8 Effects on Nitrogen Assimilation

Very few studies exist on the ways in which lead affects nitrogen assimilation. Nitrate reductase, the rate-limiting enzyme in the nitrate assimilation process, is inhibited by 10–100 mM lead in soybean leaves (Venketramana et al. 1978). At 100 mM, lead inhibited nitrate uptake and *in vivo* nitrate reductase enzyme activity in cucumber seedlings (Burzynski and Grabowski 1984). The activity of nitrate reductase was also inhibited in soybean (Huang et al. 1974), *Zostera marina* roots (Brackup and Capone 1985), *Zea mays* and *Pisum sativum* leaves (Sinha et al. 1988a,b), and *Sesamum indicum* (Kumar et al. 1992) (see Table 1). Sinha et al. (1994) reported an 81% –87% decline in nitrate reductase activity in the presence of 0.1 or 0.5 mM lead acetate. The inhibition may result from oxidation of NADH by lead or from reduced NADH production caused by mitochondrial swelling (Gangeubach et al. 1973). Alternatively, affects on nitrate reductase activity may result from reduction of the $NO_3^-$ supply to the site of enzyme synthesis because lead treatment can create water stress to the plants (Shaner and Boyer 1976;

Burzynski and Jacob 1983; Burzynski and Grabowski 1984). Glutamine synthetase activity in soybean leaves is inhibited by 100 ppm lead (Lee et al. 1976). There is also a concentration-dependent inhibition of glutamine synthetase activity in excised maize leaves (Sinha et al. 1994). Lead may interfere with de novo synthesis of the enzyme or bind to enzyme functional –SH groups (Prassad and Prassad 1987). Lead inhibited soybean growth and the associated nitrogen-fixing nodules inhabited by *Rhizobium japonicum* S-24 (Huang et al. 1974). Nodulation in soybean was inhibited at a 1.0 mM lead concentration (Paivoke 1983). It is suggested that lead diminishes the capacity of soybeans to fix nitrogen by affecting nitrogenase activity, possibly through negative regulation by the nif gene product as found recently in *Klebsiella pneumoniae* (Handerson et al. 1989).

## 3.9 Pigment Control

Changes in natural coloration are often visible signs of lead toxicity to plants. Therefore, many investigators have examined the effect of lead on pigment content, especially chlorophyll levels. Lead exposure decreases chlorophyll content in oats (Fiussello and Molinari 1973), cucumber (Burzynski 1985), dodder (Jana et al. 1987), mungbean (Prassad and Prassad 1987), and maize (Sinha et al. 1994). Parekh and Puranik (1992) also observed a decrease in chlorophyll content after application of 0.001–0.5 mM solutions of lead acetate to maize leaf segments.

It has been observed that lead inhibits δ-amino levulinic acid dehydratase activity (Burzynski 1985; Prassad and Prassad 1987; Parekh and Puranik 1992), the enzyme responsible for the synthesis of δ-amino levulinic acid, an important intermediate in chlorophyll biosynthesis. Burzynski (1985) has suggested that the reduction in chlorophyll biosynthesis by lead may result from changes in tissue hydration, which is decreased by about 50% compared to the control. The pigment is affected by lead in mungbean as well (Kumar 1983).

## 4 Summary

Lead is a metallic pollutant emanating from various environmental sources including industrial wastes, combustion of fossil fuels, and use of agrochemicals. Lead may exist in the atmosphere as dusts, fumes, mists, and vapors, and in soil as a mineral. Soils along roadsides are rich in lead because vehicles burn leaded gasoline, which contributes to environmental lead pollution. Other important sources of lead pollution are geological weathering, industrial processing of ores and minerals, leaching of lead from solid wastes, and animal and human excreta. Lead is nondegradable, readily enters the food chain, and can subsequently endanger human and animal health. Lead is one of the most important environment pollutants and deserves the increasing attention it has received in recent decades.

The present effort was undertaken to review lead stress effects on the physiobiochemical activity of higher plants. Lead has gained considerable attention as a potent heavy metal pollutant because of growing anthropogenic pressure on the environment. Lead-contaminated soils show a sharp decline in crop productivity. Lead is absorbed by plants mainly through the root system and in minor amounts through the leaves. Within the plants, lead accumulates primarily in roots, but some is translocated to aerial plant parts. Soil pH, soil particle size, cation-exchange capacity, as well as root surface area, root exudation, and mycorrhizal transpiration rate affect the availability and uptake of lead by plants. Only a limited amount of lead is translocated from roots to other organs because there are natural plant barriers in the root endodermis. At lethal concentrations, this barrier is broken and lead may enter vascular tissues. Lead in plants may form deposits of various sizes, present mainly in intercellular spaces, cell walls, and vacuoles. Small deposits of this metal are also seen in the endoplasmic reticulum, dictyosome, and dictyosome-derived vesicles.

After entering the cells, lead inhibits activities of many enzymes, upsets mineral nutrition and water balance, changes the hormonal status, and affects membrane structure and permeability. Visual, nonspecific symptoms of lead toxicity are stunted growth, chlorosis, and blackening of the root system. In most cases, lead inhibition of enzyme activities results from the interaction of the metal with enzyme –SH groups. The activities of metalloenzymes may decline as a consequence of displacement of an essential metal by lead from the active sites of the enzymes. Lead decreases the photosynthetic rate of plants by distorting chloroplast ultrastructure, diminishing chlorophyll synthesis, obstructing electron transport, and inhibiting activities of Calvin cycle enzymes.

# References

Ahmed A, Tajmir-Riahi HA (1993) Interaction of toxic metal ions $Cd^{2+}$, $Hg^{2+}$ and Pb with light-harvesting proteins of chloroplast thylakoid membranes. An FTIR spectroscopic study. J Inorg Biochem 50:235–243.

Angelone M, Bini C (1992) Trace element concentrations in soils and plants of western Europe. In: Adriano DC (ed) Biogeochemistry of Trace Metals. Lewis, Boca Raton, London, pp 19–60.

Baker WG (1972) Toxicity levels of mercury, lead, copper and zinc in tissue culture systems of cauliflower lettuce, potato and carrot. Can J Bot 50:973–976.

Bala R, Setia RC (1990) Some aspects of Cd and Pb toxicity in plants. Adv Front Areas Plant Sci 19:167–180.

Baumhardt GR, Welch LF (1972) Lead uptake and corn growth with soil applied lead. J Environ Qual 1:92–94.

Bazzaz FA, Carlson RA, Rolfe BL (1974) The effect of heavy metals on plant I. Inhibition of gas exchange in sunflower by lead, cadmium nickel and thallium. Environ Pollut 7:241–246.

Bazzaz FA, Carlson RW, Rolfe GL (1975) The inhibition of corn and soybean photosynthesis by lead. Physiol Plant 34:26–329.

Bazzaz MB, Govindji (1974) Effect of lead chloride on chloroplast reactions. Environ Lett 6175:191.

Blaylock MJ, Salt DE, Dushenkov S, Zakarova O, Gussman C, Kapulnik Y, Ensley BD, Raskin I (1997) Enhanced accumulation of Pb in Indian mustard by soil-applied chelating agents. Environ Sci Technol 31 860–865.

Brackup I, Capone DG (1985) The effect of several metal and organic pollutants on nitrogen fixation (acetylene reduction) by the roots and rhizome of *Zostera marina* L. Environ Exp Bot 25:145–151.

Bradshaw AD (1952) Population of *Agrostis tenuis* resistant to lead and zinc poisioning. Nature (Lond) 169:1098.

Bradshaw AD, McNeilly T(1981) Evolution and Pollution. Studies in Biology, vol 130. Arnold, London.

Broyer R, Johnson CM, Paull RE (1972) Some aspects of lead in plant nutrition. Plant Soil 36:301–313.

Burzynski M (1985) Influence of lead in chlorophyll content and on initial step of its synthesis in greening cucumber (*Cucumis sativa*) seedlings. Acta Soc Bot Pol 54:95–106.

Burzynski M (1987) The influence of lead and cadmium on the absorption and distribution of potassium, calcium, magnesium and iron in cucumber seedlings. Acta Physiol Plant 9:229–238.

Burzynski M, Grabowski A (1984) Influence of lead on nitrate uptake and reduction in cucumber seedlings. Acta Soc Bot Pol 53:77–86.

Burzynski M, Jakob M (1983) Influence of lead on auxin-induced cell elongation. Acta Soc Bot Pol 52:231–239.

Canon HL, Bowler JM (1962) Contamination of vegetation of tetraethyl lead. Science 137:765–766.

Carlson RW, Bazzaz FA (1977) Growth reduction in American sycamore (*Platanus occidentalis*) caused by lead-cadmium interaction. Environ Pollut 12:243–253.

Chaney RL, Ryan JA (1994) Risk Based Standards for Arsenic, Lead and Cadmium in Urban Soils. Dechema, Frankfurt, Germany.

Chow TJ (1970) Lead accumulation in roadside soil and grass. Nature (Lond) 225:295.

Cox WJ, Rains DW (1972) Effect of time on lead uptake by five plant species. J Environ Qual 1:167–169.

Daines RH, Motto HL, Chilcko DM (1970) Atmospheric lead: its relationship to traffic volume and proximity to highways. Environ Sci Technol 4:318–322.

Davies BE (1995) Lead and other heavy metals in urban areas and consequences for the health of their inhabitants. In: Majumdar SK, Miller EW, Brenner FJ (eds) Environmental Contaminants, Ecosystems and Human Health. The Pennsylvania Academy of Science, Easton, PA, pp 287–307.

de Abreu CA, de Abreu MF, de Andrade JC (1998) Distribution of lead in the soil profile evaluated by DTPA and Mehlich-3 solutions. Bragantia 57:185–192.

De Treville F (1964) Natural occurrence of lead. Arch Environ Health 8:212–221.

De Filippis LF, Pallaghy CK (1976) The effect of sublethal concentration of mercury, and zinc on *Chlorella*. II. Photosynthesis and pigment composition. Z Pflanzenphysiol 78:314–322.

De Filippis LF, Hampp R, Ziegler H (1981) The effects of sublethal concentrations of zinc, cadmium and mercury on *Euglena*: respiration, photosynthesis and photochemical activities. Arch Microbiol 128:407–411.

Denny P, Weeks DC (1968) Electropotential gradients of ions in an aquatic angiosperm *Potamegeton schweinfurthii* (Benn). New Phytol 67:875–882.

Drazkiewicz M (1994) Chlorophyll occurrence, functions, mechanism of action, effects of internal and external factors. Photosynthetica 30:321–331.

Eick MJ, Peak JD, Brady PV, Pasak JD (1999) Kinetics of lead adsorption and desorption on goethite: residence time effect. Soil Sci 164:28–39.

Fiussello N, Molinari MT (1973) Effect of lead on plant growth. Allionia 19:86–96.

Foy CD, Chaney RL, White MC (1978) The physiology of metal toxicity in plants. Annu Rev Plant Physiol 29:511–566.

Gangeubach BG, Miller RJ, Koepple DE, Arntzen CJ (1973) The effect of toxin from *Helminthosporium maydis* (race 1) on isolated corn mitochondria swelling. Can J Bot 51:2119–2125.

Glater RA, Hernandez L (1972) Lead detection in living plant tissue using a new biochemical method. J Air Pollut Control Assoc 22:463–467.
Godzik B (1993) Heavy metal contents in plants from zinc dumps and reference area. Pol Bot Stud 5:113–132.
Goldbold DJ, Hutterman A (1986) The uptake and toxicity of mercury and lead to spruce (*Picea abies*) seedlings. Water Air Soil Pollut 31:509–515.
Goldsmith CD Jr, Scalon PF, Pirie WF (1976) Lead concentration soil and vegetation associated with highways of different traffic densities. Bull Environ Contam Toxicol 16:66–70.
Gruenhage L, Jager HJ (1985) Effect of heavy metals on growth and heavy metal content of *Allium porrum* and *Pisum sativum*. Angew Bot 59:11–28.
Hampp R, Zeigler H, Zeigler I (1973) Influence of lead ions on the activity of enzymes of reductive pentose phosphate pathway. Biochem Physiol Pflanz 164:588–495.
Handerson H, Asutin S, Dixon RA (1989) Role of metal ions in negative regulation of nitrogen fixation by the nif L. gene product from *Klebsiella pheumoniae*. Mol Genet 216:484–491.
Hardimann RT, Jacob B, Banin A (1984) Factors affecting the distribution of cadmium, copper and lead and their effect upon yield and zinc content in bush bean (*Phaseolus vulgaris*). Plant Soil 81:17–27.
Hewitt EJ (1953) Metal inter-relationships in plant nutrition. J Exp Bot 4:59–64.
Hinsely TD, Bradis D, Malina JE (1971) Agriculture benefits and environmental changes resulting from the use of digested sewage sludge on field crops. USEPA Solid Waste Demonstration Grant GOE-EC-0080, Cincinnati, OH, 1339:122–124.
Huang CV, Bazzaz FA, Venderhoef LN (1974) The inhibition of soybean metabolism by cadmium and lead. Plant Physiol 34:122–124.
Huang JW, Cunningham SD (1996) Lead phytoextraction species variation in lead uptake and translocation. New Phytol 134:75–84.
Huang JW, Grumes DL, Kochian LV (1994) Voltage dependent $Ca^{++}$ influx into right-side-out plasma membrane vesicles isolated from wheat roots: characteristics of a putative $Ca^{++}$ channel. Proc Natl Acad Sci U S A 91:3473–3477.
Iqbal J, Mushtaq S (1987) Effect of lead on germination, early seedling growth, soluble protein and acid phosphatase content in *Zea mays*. Pak J Sci Ind Res 30:853–856.
Jana S, Choudhuri MA (1984) Synergistic effect of heavy metal pollutants on senescence in submerged aquatic plants. Water Air Soil Pollut 21:351–357.
Jana S, Dalal T, Barua B (1987) Effects and relative toxicity of heavy metals of *Cascuta reflexa*. Water Air Soil Pollut 33:23–27.
Jarvis MD, Leung DWM (2002) Chelated lead transport in *Pinus radiata*: an ultrastructural study. Environ Exp Bot 48:21–32.
Jawarski JF (1978) Effect of Lead in the Environment. Qualitative Aspects. Publication no. NRCC.16736 of Environmental Secretariate, BRCC Publication, Ottawa.
John MK, Van Laerhovexn C (1972) Lead uptake by lettuce and oats as affected by time, nitrogen and source of lead. J Environ Qual 1:169–171.
Johnson MS, Eaton JW (1980) Environmental contamination through residual trace metal dispersal from a derelict lead-zinc mine. J Environ Qual 9:175–179.
Johnson MS, McNeilly T, Putwain PO (1977) Revegetation of metalliferous mine spoil contaminated by lead and zinc. Environ Pollut 12:261–277.
Johnson WR, Proctor J (1977) A comparative study of metal levels in plants from two contrasting lead mine sites. Plant Soil 46:251–257.
Jones LHP, Clement CR, Hooper MS (1973) Lead uptake and its transport from roots to shoots. Plant Soil 38:403–414.
Juwarkar AS, Shende GB (1986) Interaction of Cd-Pb effect on growth yield and content of Cd, Pb in barley. Indian J Environ Health 28:235–243.
Kaji T, Fujiwara Y, Yamamoto C, Sakamoto, Kozuka H (1995) Inhibitory effect of lead on the proliferation of culture vascular endothelial cells. Toxicology 95(1–3):87–92.
Karawmanos RE, Battany JR, Stewart JWB (1976) The uptake of native and applied lead by alfalfa and bromegrass from soil. Can Sci 56:485–494.

Keaton CM (1937) The influence of Pb compounds on growth of barley. Soil Sci 43:401–409.
Keller T, Zuber R (1970) Lead uptake and distribution in young spruce plants. Forstwiss Centrabl (Hamb) 89:20–26.
Khan S, Khan NN (1983) Influence of lead and cadmium on growth and nutrient concentration of tomato (*Lycopersicum esculentum*) and eggplant (*Solanum melongera*). Plant Soil 74:387–394.
Koeppe DE (1977) The uptake, distribution and effect of cadmium and lead in plants. Sci Total Environ 7:197–205.
Koeppe DE (1981) Lead: understanding the minimal toxicity of lead in plants. In: Lepp NW (ed) Effect of Trace Metal on Plant Function. Applied Science, London, pp 55–76.
Koeppe DE, Miller RJ (1970) Lead effects on corn mitochondrial respiration. Science 167:1376–1377.
Kosobrukhov A, Knyazeva I, Mudrik V (2004) *Plantago major* plant responses to increase content of lead in soil: growth and photosynthesis. Plant Growth Regul 42:145–151.
Krishnayapa JSR, Bedi SJ (1986) Effect of automobile lead pollution on *Cassia tora* and *Cassia occidentalis* L. Environ Pollut 40:221–226.
Kumar G (1990) Growth, photosynthetic pigments and nitrate assimilation in *Sesamum indicum* L. var. F 5-1 in a lead enriched environment. M. Philos. dissertation, M.D. University, Rohtak.
Kumar G, Singh RP, Sushila (1992) Nitrate assimilation and biomass production in *Sesamum indicum* (L.) seedlings in lead enriched environment. Water Air Soil Pollut 215:124–215.
Kumar N (1983) Effect of lead on growth and anthocyanin development of *Phaseolus auseus*. Acta Bot Indica 11:71–72.
Kumar NPBA, Dushenkov V, Motto H, Raskin I (1995) Phytoextraction: the use of plants to remove heavy metals from soils. Environ Sci Technol 29:1232–1238.
Lagerwerff JV, Armiger WH, Specht AW (1973) Uptake of lead by alfalfa and corn from soil and air. Soil Sci 115:455–460.
Lamoreaux RJ, Chaney WR (1978) The effect of cadmium on net photosynthesis, transpiration and dark respiration of excised silver maple leaves. Physiol Plant 43:231–236.
Lane SD, Martin ES (1977) A histochemical investigation of lead uptake in *Raphanus sativus*. New Phytol 79:281–286.
Lane SD, Martin ES, Ganod JR (1978) Lead toxicity effects on indole-3-acetic acid induced cell elongation. Planta 144:79–84.
Laxen DPH, Harrison RM (1977) The highway as a source of water pollution: an appraisal of heavy metal lead. Water Res 11:1–11.
Lee KC, Cunniggham BA, Chung KH, Paulson GM, Liang GH (1976) Lead effects on several enzymes and nitrogenous compounds in soybean leaves. J Environ Qual 5:357–359.
Lee S-Z, Chang L, Yang H-H, Chen C-M, Liu M-C (1998) Absorption characteristics of lead onto soils. J Hazard Mat 63:37–49.
Liebhardt NC, Koske TJ (1974) The lead content of various plant species as affected by Cycle-Lite® humus. Commun Soil Sci Plant Annu 5:357–359.
Liou S, Trong N, Horn-Chechiang W, Guang Young Y, Yea Quay W, Jini Shoung L, Tsong H, Yuchiang G, Ying-Chin K, Po-ya C (1974) Blood lead levels in the general population of Taiwan Republic of China. Int Arch Occup Environ Health 66(4):255–260.
Maier R (1977) The effect of lead on the NADH dependent malate dehydrogenase in *Medicago sativa* and *Zebrina pendula* S. Z Pflanzenphysiol 85:319–326.
Maier R (1978) Studies on the effect on lead on the acid phosphatase in *Zea mays* L. Z Pflanzenphysiol 87:347–354.
Malone CD, Koeppe DE, Miller RJ (1974) Localization of lead accumulated by corn plants. Plant Physiol 536:388–394.
Marschner P, Godbold DL, Jutschhe G (1996) Dynamics of lead accumulation in mycorrhizal and non-mycorrhizal Norway spruce (*Picea abies* (L.)) karst. Plant Soil 178:239–245.
Marshall J, Corzo A, Leigh RA, Sanders D (1994) Membrane potential-dependent calcium transport in right-side-out plasma membrane vesicles from *Zea mays* L. roots. Plant J 5:683–694.

Martin GC, Hammond PB (1966) Lead uptake by brome grass from contaminated soil. Agron J 58:553–554.

Miles CD, Brandle JR, Daniel DJ, Cheu D, Schnore PD, Unlik PJ (1972) Inhibition of photosystem II in isolated chloroplasts by lead. Plant Physiol 49:820–825.

Miller JE, Hassete JJ, Koppe DE (1975) Interaction of lead and cadmium on metal uptake and growth of corn plants. J Environ Qual 6:18–20.

Miller RJ, Biuell JE, Koeppe DE (1973) The effect of cadmium on electron and energy transfer reactions in corn mitochondria. Physiol Plant 28:166–171.

Mohanty N, Vass I, Demeter S (1989) Copper toxicity effects: photosystem II electron transport at the secondary quinone acceptor, $Q_B$. Plant Physiol 90:175–179.

Morzek E Jr, Funicelli NA (1982) Effect of lead and zinc on germination of *Spartina alterniflora* Loisel seeds at various salinities. Environ Exp Bot 22:23–32.

Mukherji S, Maitra P (1976) Toxic effects of lead on growth and metabolism of germinating rice (*Oryza sativa* L.) seeds on mitosis of onion (*Allium cepa*) root tip cells. Indian J Exp Biol 14:519–521.

Nakos G (1979) Lead pollution. Fate of lead in the soil and its effects on *Pinus haplensis*. Plant Soil 53:427–473.

Niragh JO (1988) A silent epidemic of environmental metal poisioning. Environ Pollut 50:159–161.

Pages AL, Ganje TJ (1970) Accumulation of lead in soils for regions of high and low motor vehicle traffic density. Environ Sci Technol 4:307–315.

Paivoke AEA (2002) Soil lead alters phytase activity and mineral nutrient balance of *Pisum sativum*. Environ Exp Bot 48:61–73.

Paivoke H (1983) The short term effect of zinc on the growth anatomy and acid phosphatase activity of pea seedlings. Ann Bot 20:307–309.

Parekh D, Puranik RM (1992) Inhibition of chlorophyll biosynthesis lead in greening maize leaf segments. Indian J Exp Biol 30:302–304.

Parys E, Romanowaska E, Siedlecka M, Poskuta J (1998) The effect of lead on photosynthesis and respiration in detached leaves in mesophyll protoplasts of *Pisum sativum*. Acta Physiol Plant 20:313–322.

Patel PM, Wallace A, Romney EM (1977) Effect of chelating agents on phototoxicity of lead and lead transport. Commun Soil Sci Plant Annu 8:733–740.

Pilgrim W, Hughes RN (1994) Lead, cadmium, arsenic and zinc in the ecosystem surrounding a lead smelter. Environ Monit Assess 32(1):1–20.

Porter JR, Cheridan RP (1981) Inhibition of nitrogen fixation in alfalfa by arsenate, heavy metals, fluoride, and simulated acid rain. Plant Physiol 68:143–148.

Prassad DDK, Prassad ARK (1987) Altered δ-aminolaevulinic acid metabolism by lead and mercury in germinating seedlings of Bajra (*Pennisetum typhoideum*). J Plant Physiol 127:241–249.

Purves D (1977) Trace Element Contamination of the Environment, vol 1. Elsevier, Amsterdam.

Rashid A, Camm EL, Ekramoddoullah KM (1994) Molecular mechanism of action of Pb and $Zn^{2+}$ on water oxidizing complex of photosystem II. FEBS Lett 350:296–298.

Rashid P, Mukherji S (1991) Changes in catalase and ascorbic oxidase activities in response to lead nitrate treatments in mungbean. Indian J Plant Physiol 34:143–146.

Rashid P, Mukerji S (1993) Effect of foliar application of lead on growth and yield parameter of wheat. Pak J Sci Ind Res 36 (11):473–475.

Rebechini HM, Hanzely L (1974) Lead-induced ultrastructural changes in chloroplasts of the hydrophyte *Ceratophyllum demersum*. Z Pflanzenphysiol 73:377–386.

Reese RN, Roberts LW (1985) Effects of cadmium on whole cell and mitochondrial respiration in tobacco cell suspension cultures (*Nicotiana tobacum* L. var. *xanthi*). J Plant Physiol 120:123–130.

Roderer G (1984) On the toxic effect of tetraethyl lead and its derivatives on the chryosophyte *Poterioochromonas malhamensis*. V. Electron microscopical studies. Environ Exp Bot 24:17–30.

Rolfe GL, Bazzaz FA (1975) Effect of lead contamination on transpiration and photosynthesis of loblolly pine and autumn olive. For Sci 21:33–35.

Romanowska E, Igamberdiev AU, Parys E, Gardeström P (2002) Stimulation of respiration by Pb in detached leaves and mitochondria of $C^3$ and $C^4$ plants. Physiol Plant 116:148–154.

Rudakova EV, Karakis KD, Sidorshina ET (1988) The role of plant cell walls in the uptake and accumulation of metal ions. Phyziol Biochim Kult Rast 20:3–12.

Salim RM, Isa MM, Al Subu, Sayrafi SA, Sayrafi O (1995) Effect of irrigation with lead and cadmium on the growth and metal uptake of cauliflower, spinach and parsley. J Environ Sci Health Part A 30(4):831–849.

Samardakiewicz S, Wozny A (2000) The distribution of lead in duckweed (*Lemna miner* L.) root tip. Plant Soil 226:107–111.

Sarkar A, Jana S (1987) Effect of combination of heavy metals on Hill activity of *Azolla pinnata*. Water Air Soil Pollut 35:141–145.

Sarvari E, Gaspar L, Fodor F, Cseh E, Kropfl K, Varga A, Barnon M (2002) Comparison of the effects of Pb treatment on thylakoid development in poplar and cucumber plants. Acta Biol Szeged 46:163–165.

Schmidt GH, Ibrahim MM (1994) Heavy metal content ($Hg^{2+}$, $Cd^{2+}$, $Pb^{2+}$) in various body parts, its impact on choline esterase activity and binding glycorproteins in grasshopper (*Aeolopus thalassins*) adults. Ecotoxicol Environ Saf 29(2):148–164.

Sengar RS, Pandey M (1996) Inhibition of chlorophyll biosynthesis by lead in greening *Pisum sativum* leaf segments. Biol Plant 38:459–462.

Seregin IV, Shpigun LK, Ivaniov VB (2004) Distribution and toxic effects of cadmium and lead on maize roots. Russ J Plant Physiol 51:525–533.

Sereson F, Kralova K, Bumbalova A (1998) Action of mercury on the photosynthetic apparatus of spinach chloroplasts. Photosynthetica 35:551–559.

Shaner DL, Boyer JS (1976) Nitrate reductase activity in maize (*Zea mays* L.) leaves. Plant Physiol 58:499–504.

Sieghardt H (1984) An anatomical, histochemical investigation on the distribution of lead in primary roots of *Pisum sativum* L. Mikroskopie (Wien) 41:125–133.

Silverberg BA (1975) Ultrastructural localization of lead in *Stigeoclonium tenue* (Chlorophyseae: Ulotrichales) as demonstrated by cytochemical and x-ray microanalysis. Phycologia 14:265–274.

Sinha SK, Srivastava HS, Mishra SN (1988a) Nitrate assimilation in intact and excised maize leaves in the presence of lead. Bull Environ Contam Toxicol 41:419–42.

Sinha SK, Srivasatava HS, Mishra SN (1988b) Effect of lead on nitrate reductase activity and nitrate assimilation in pea leaves. Bot Pol 57:457–463.

Sinha SK, Srivastava HS, Tripathi TD (1994) Influence of some growth regulators and divalent cations on the inhibition of nitrate reductase activity by lead in maize leaves. Chemosphere 29:1775–1782.

Smith WH (1972) Lead and mercury burden of urban woody plants. Science 176:1237–1238.

Stefanov K, Seizova K, Popova I, Petkov VL, Kimenov G, Popov S (1995) Effects of lead ions on the phospholipid composition in leaves of *Zea mays* and *Phaseolus vulgaris*. J Plant Physiol 147:243–246.

Stiborova M, Pitrichova M, Brezinova A (1987) Effect of heavy metal ions on growth and biochemical characterstic of photosynthesis of barley and maize seedlings. Biol Plant 29:453–467.

Strand V, Zolotareva BN, Lisovskij AJ (1990) Effect of lead, cadmium and copper content in soil on their accumulation, in and yields of crops. Rosthinna 36(4):411–417.

Suchodoller A (1967) Untersuchungen ueber den Bleigehatt von Pflanzen in der Nane von Strassen und ueber die Aufnahme and Translokation von Blei durch Pflazen. Ber Schweiz Bot Ges 77:266–308.

Sudhakar C, Syamalabai L, Veeranjaveyuler K (1992) Lead tolerance of certain legume species grown on lead or tailing. Agric Ecosyst Environ 41(3-4):253–261.

Symeonidis SL, McNeilly J, Bradshaw AD (1985) Differential tolerance of *Agrostis capillaris* to cadmium copper, lead, nickel and zinc. New Phytol 101:309–316.

Tandler CJ, Solavi AJ (1969) Nucleolar orthophosphate ion electron microscope and diffraction studies. J Cell Biol 41:91–108.

Taylor TW, Allinson DW (1981) Influence of lead, cadmium, nickel on growth of *Medicago sativa*. Plant Soil 60:223–236.

Tomsig JL, Suszkiw JB (1991) Permeation of Pb through calcium channels: fura-2 measments of voltage-and dihydropyridine-sensitive Pb entry in isolated borine chromaffin cells. Biochim Biophys Acta 1069:197–200.

Valle BL, Ulmer DD (1972) Biochemical effect of mercury, cadmium and lead. Annu Rev Biochem 41:91–128.

Van Assche F, Clijesters H (1990) Effect of metals on enzyme activity in plants. Plant Cell Environ 13:95–206.

Venketramana S, Veeranjaneyulu K, Rama Das VS (1978) Inhibition of nitrate reductase by heavy metal. Indian J Exp Biol 16:615.

Verma S, Dubey RS (2003) Lead toxicity induces lipid peroxidation and alters the activities of antioxidant enzymes in growing rice plants. Plant Sci 164:645–655.

Vodnik D, Jentschke G, Fritz E, Gogala N, Godbold DL (1999) Root-applied cytokinin reduces lead uptake and affects its distribution in Norway spruce seedlings. Physiol Plant 106:75–81.

Wallace A, Rombey EM, Alexander GV, Soufi SM, Patel PM (1976) Some interaction in plants among cadmium, other heavy metals and chelating agents. Agron J 69:120–123.

Wheeler GL, Rolfe GL (1979) The relationship between daily traffic volume and the distribution of lead in roadside soil and vegetation. Environ Pollut 18:265–274.

Wierzbicka M (1987) Lead accumulation and its translocation in roots of *Allium cepa* L.: autoradiographic and ultrastructural studies. Plant Cell Environ 10:17–26.

Wierzbicka M (1994) Resumption of mitotic activity in *Allium cepa* root tips during treatment with lead salts. Environ Exp Bot 34:173–180.

Wierzbicka M (1998) Lead in the apoplast of *Allium cepa* L. root tips: ultrastructural studies. Plant Sci 133:105–119.

Wozny A, Raman P, Mlodzianowski F (1982) The effect of kinetin on cytochemical localization of magnesium dependent ATPase in isolated lupin cotyledons. Acta Soc Bot Pol 5:345.

Wright JR, Levic R, Atkinson HJ (1995) Trace element distribution in virgin profiles representing four great soil groups. Soil Sci Am Proc 19:340–344.

Yang Y-Y, Jung J-Y, Song W-Y, Suh HS, Lee Y (2000) Identification of rice varieties with high tolerance or sensitivity to lead and characterization of the mechanism of tolerance. Plant Physiol 124:1019–1026.

Zimdahl RL (1976) Entry and movement in vegetation of lead derived from air and soil sources. Am. Pollut Cont Assoc J 7:655–659.

Zimdahl RL, Faster JM (1976) The influence of applied phosphorus manure or lime on uptake of lead from soil. J Environ Qual 5:31–34.

Zimdahl RL, Koeppe DE (1977) Uptake by plants. In: Boggess WR (ed) Lead in the Environment. National Science Foundation, Washington, DC, pp 99–104.

# Environmental Fate and Toxicology of Carbaryl

Amrith S. Gunasekara, Andrew L. Rubin, Kean S. Goh,
Frank C. Spurlock, and Ronald S. Tjeerdema

### Contents

1 Introduction .................................................................... 95
2 Chemistry ...................................................................... 96
3 Chemodynamics ............................................................. 97
   3.1 Air ........................................................................... 97
   3.2 Water ...................................................................... 98
   3.3 Soil ......................................................................... 98
4 Degradation ................................................................... 101
   4.1 Abiotic .................................................................... 101
   4.2 Biotic ...................................................................... 104
5 Toxicity ......................................................................... 107
   5.1 Insects and Aquatic Organisms ............................. 107
   5.2 Animals .................................................................. 108
   5.3 Humans .................................................................. 112
6 Mammalian Toxicokinetics and Metabolism ................. 113
7 Summary ....................................................................... 115
References ........................................................................ 116

## 1 Introduction

Carbaryl (1-naphthyl-*N*-methyl carbamate; Fig. 1), a carbamate insecticide introduced in 1956 by Union Carbide Corporation, is used worldwide and is a substitute for some organochlorine insecticides (Ribera et al. 2001). Carbaryl is used to

---

A.S. Gunasekara[1], R.S. Tjeerdema
Department of Environmental Toxicology, College of Agricultural and Environmental Sciences, University of California, Davis, CA 95616-8588, USA

A.L. Rubin, K.S. Goh, F.C. Spurlock
Department of Pesticide Regulation, California Environmental Protection Agency, Sacramento, CA 95812-4015, USA

[1] Present address: Division of Inspection Services (FFLDERS), California Department of Food and Agriculture, 1220 N Street, Sacramento, CA 95814, USA
e-mail: agunasekara@cdfa.ca.gov

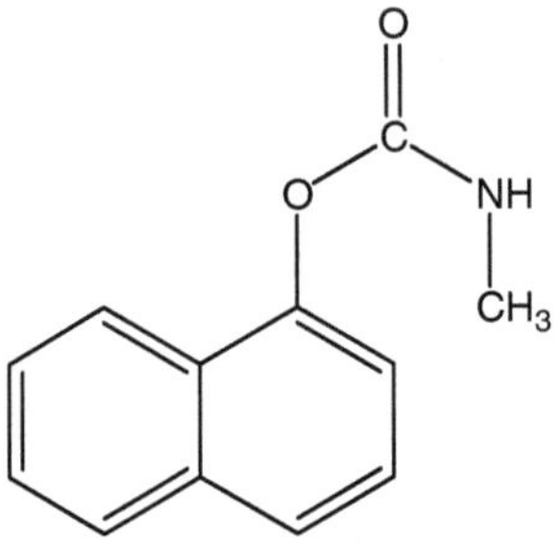

**Fig. 1** Chemical structure of carbaryl

control a broad spectrum of insects on more than 120 different crops (Ware 2000). It has also been used to prevent bark beetle infestation in pine trees (Hastings et al. 2001) and as a general garden insecticide (Ware 2000). In 2005, approximately 189,800 lbs of the insecticide was applied in California alone (CDPR 2005). Annual use in the United States has been reported to be 4.5–6.8 million kg (Cox 1993). Several trade names are associated with carbaryl (the most common is Sevin®), and active ingredient (a.i.) use rates range from 0.57 to 4.5 kg/ha (Rajagopal et al. 1984). Carbaryl is available in the forms of a wettable powder, pellets, granules, suspensions, and solutions, and is the second most widely detected insecticide in surface waters in the U.S. (Martin et al. 2003).

## 2 Chemistry

Carbaryl, similar to most carbamates, inhibits the enzyme acetylcholinesterase (AChE), which is responsible for the degradation of the neurotransmitter acetylcholine in insects. Its inhibition promotes the buildup of AChE at synaptic junctions, resulting in uncontrolled movement, paralysis, convulsions, and possible death (Tomlin 2000). AChE inhibition also causes the toxicity of carbaryl to mammals, although, in contrast to insects, the mammalian effect involves synapses in the peripheral nervous system, including those in glandular structures and at neuromuscular junctions, in addition to those in the central nervous system. Because of the hydrolytic instability of the carbamate-AChE bond, recovery of mammals from acute effects is expected when exposures are low. Other cholinesterases (ChEs) inhibited by carbaryl include the plasma-localized butyryl ChE and the red blood cell-localized AChE. Evidence for inhibition of plasma and/or red blood cell ChEs can be interpreted and used as an indicator of exposure. The physicochemical properties of carbaryl are listed in Table 1; it has a low molecular weight, is moderately soluble in water, and does not readily volatilize (Tomlin 2000).

**Table 1** Physicochemical properties of carbaryl

| Pure physical state[a] | | Colorless or tan crystal |
|---|---|---|
| Chemistry Abstracts Service registry number (CAS #)[b] | | 63-25-2 |
| Molecular weight (g/mol)[a] | | 201.2 |
| Molecular formula[a] | | $C_{12}H_{11}NO_2$ |
| Melting point (°C)[a] | | 142 |
| Vapor pressure (mPa at 23.5°C)[a] | | 0.041 |
| Octanol–water partition coefficient (log $K_{ow}$)[a] | | 2.36 |
| Density (20°C)[a] | | 1.23 |
| Henry's law constant (atm m³ g/mol at 25°C)[a] | | $2.74 \times 10^{-9}$ |
| Organic-carbon normalized partition coefficient ($K_{oc}$)[b] | | 290 |
| $\lambda_{max}$ (nm)[c] | | 280 |
| Water solubility (mg/L) | 20°C[a] | 120 |
| | 25°C[d] | 104 |
| | 40°C[e] | 40 |

*Sources*: [a]Tomlin (2003); [b]Phillips and Bode (2004); [c]Sheng et al. (2001); [d]Arroyo et al. (2004); [e]Meister (2001).

# 3 Chemodynamics

## 3.1 Air

Carbaryl has low volatility because of its low vapor pressure (see Table 1). Additionally, its low Henry's law constant suggests that it will not volatilize from aqueous solutions (Table 1). However, carbaryl may become airborne from sorption to particulates or as a spray drift immediately following application. Drift monitoring from aerial spraying at a rate of 2,250 g a.i./ha on a Vermont apple orchard showed concentrations of 0.70–7.20 μg/plate (a 1-mm-thick Teflon sheet covered the 15-cm-diameter Petri plate), which corresponds to 0.4–4.1 g a.i./ha, as far out as 305 m with 8–12 km/hr winds (Currier et al. 1982). Higher concentrations were observed at 76 m downwind (481 μg/plate) and 12 m upwind (45.9 μg/plate) in the same study. However, it was also noted that all detections decayed to relatively low concentrations within 2 hr after application (<2 ug/m³; Currier et al. 1982). Airborne carbaryl degrades after reaction with photochemically produced hydroxyl radicals in the atmosphere (Kao 1994), with a reaction rate constant of $3.3 \times 10^{-11}$ cm³/sec (Sun et al. 2005).

Low drift concentrations were reported in a California study, with concentrations of up to 1.12 μg/m³ in the air after ground spraying to control the glassy-winged sharpshooter, *Homalodisca coagulate* (Walters et al. 2003). Although below the adverse health effect concentration (51.7 μg/m³), the insecticide was present in air up to 47 hr after application (Walters et al. 2003). Shehata et al. (1984) reported atmospheric concentrations of some 0.0035–0.107 μg/m³ over a Maine forest treated with carbaryl to control the spruce budworm.

In eastern France, atmospheric measurements for carbaryl at remote (non-populated), rural (population, 80,000), and urban (population, 300,000) sites were, on average, 280, 348, and 577 pg/m$^3$, with highest detections at 1,800, 696, and 1,420 pg/m$^3$, respectively (Sanusi et al. 2000). The increased urban and rural concentrations were mainly caused by local agricultural use (Sanusi et al. 2000). Similar concentrations were observed in 1995 at three urban and agricultural sites along the Mississippi River (Foreman et al. 2000). However, the insecticide was detected more frequently in urban versus agricultural sites in Mississippi and Iowa, possibly a reflection of its growing domestic use (Foreman et al. 2000).

## 3.2 Water

The presence of carbaryl in aquatic systems has important implications for both human and animal health because of exposure via drinking water. Carbaryl is moderately soluble in water, and its solubility predictably increases with temperature and organic solvents (see Table 1). Residues, at low (μg/L) concentrations, have been detected in surface waters adjacent to both agricultural and urban areas of some 42 states (Table 2), although several states have reported a higher frequency of detections in urban versus agricultural areas (Table 2). Carbaryl was one of the four insecticides most commonly detected in urban streams in 2001 (Gilliom et al. 2007). In Florida, Wilson and Foos (2006) reported carbaryl at 0.33–0.95 μg/L in 8 of 457 samples collected from Ten Mile Creek (an important agricultural drainage). Higher concentrations (6.94–1737 μg/L) were detected in several central California locations following carbaryl use to control the glassy-winged sharpshooter (*H. coagulate*; Walters et al. 2003). Conversely, lower concentrations (10–100 ng/L) were reported in the Pinios River of Greece following its seasonal use in the agriculturally important Thessaly region (Fytianos et al. 2006).

Carbaryl has also been found in the groundwater of several states, although at low concentrations; New Jersey reported the highest number of detections across all land-use types (see Table 2). LaFleur (1976) found carbaryl in groundwater within 2 mon of application to Congaree soil (a well-drained riverbed loam), with detection continuing up to 8 mon.

## 3.3 Soil

Sorption to soils, in general, may prevent contamination of both surface waters and groundwater by carbaryl. Soil sorption is rapid, ranging from 0.5 hr (Ahmad et al. 2001a) to 3 hr (Jana and Das 1997), but persistent (from 2 to 16 wk) with a $t_{1/2}$ of ~8 d for concentrations ranging from 1 to 14 mg/L (Rajagopal et al. 1984). Carbaryl was found to sorb more readily to acidic soils (Rajagopal et al. 1984),

**Table 2** Detection of carbaryl in U.S. surface waters and groundwaters[a]

| | | Carbaryl | | |
| --- | --- | --- | --- | --- |
| State | Type of land use | Surface water detections (no.) | Groundwater detections (no.) | Concentration range (μg/L) |
| Alabama | Urban | 61 | 1 | 0.002–0.422 |
| | Agriculture | 19 | 2 | |
| | Mixed | 41 | 1 | |
| California | Urban | 166 | – | 0.0005–5.20 |
| | Agriculture | 251 | 1 | |
| | Mixed | 432 | 1 | |
| Colorado | Urban | 190 | – | 0.0005–16.5 |
| | Agriculture | 27 | – | |
| | Mixed | 126 | 3 | |
| Florida | Urban | 39 | – | 0.003–0.441 |
| | Agriculture | 21 | – | |
| | Mixed | 39 | – | |
| Georgia | Urban | 208 | 1 | 0.001–1.90 |
| | Agriculture | 20 | – | |
| | Mixed | 177 | – | |
| Hawaii | Mixed | 8 | – | 0.007–0.370 |
| | Urban | 9 | – | |
| Indiana | Urban | 119 | – | 0.001–0.460 |
| | Agriculture | 69 | – | |
| | Mixed | 62 | – | |
| New Jersey | Urban | 122 | 5 | 0.001–1.50 |
| | Agriculture | 24 | 5 | 0.001–2.41 |
| | Mixed | 89 | 9 | |
| Pennsylvania | Urban | 119 | – | |
| | Agriculture | 82 | 1 | 0.001–5.18 |
| | Mixed | 82 | 9 | |
| Texas | Urban | 164 | 7 | |
| | Agriculture | 13 | – | 0.002–2.0 |
| | Mixed | 138 | 4 | |
| Virginia | Urban | 165 | 2 | |
| | Agriculture | 14 | – | 0.001–33.5 |
| | Mixed | 45 | 3 | |
| Washington | Urban | 46 | – | |
| | Agriculture | 267 | 1 | 0.002–0.267 |
| | Mixed | 106 | 2 | |
| Wisconsin | Urban | 27 | – | |
| | Agriculture | 8 | – | |
| | Mixed | 40 | – | |

[a] All data from USGS (2007).

and both mineral and organic fractions contributed to its sorption. Mineral interactions are clearly reported in several recent studies. For instance, Sheng et al. (2001) found that potassium-saturated smectite clay (a nonionic, expandable, hydrophilic clay) is a better sorbent for carbaryl than soil organic matter (SOM); the distribution coefficient ($K_d$) was five times greater in clay (235) than SOM-rich soil (muck; 54.2). Sheng et al. (2001) estimated that clay saturated with potassium sorbs approximately 35 times more carbaryl than a soil containing 2% SOM.

Interestingly, De Oliveira et al. (2005) found that carbaryl sorption is dependent on surface charge density and is thus site specific. For example, the amount of carbaryl sorbed was reported to be strongly dependent on the presence of specific exchangeable cations and followed the order of Ba ~ Cs ~ Ca > Mg ~ K > Na ~ Li. The polar nature of the carbonyl group was found to directly interact with exchangeable cations such as $Mg^{2+}$ and Na (De Oliveira et al. 2005). Similarly, Jana and Das (1997) demonstrated a positive correlation of carbaryl sorption with surface area, cation-exchange capacity (CEC), and free $Al_2O_3$ content in Ultisol and Inceptisol soils; sorption isotherms with Indian soils followed reversible S-shaped curves, suggesting multilayer adsorption on the sorbent surface (Jana and Das 1997).

Organic matter is another contributor to carbaryl sequestration in soils. For example, carbaryl movement through soil was found to be a function of SOM content; ~52% carbaryl was leached in 10 rinses from organic-rich soil, whereas only one rinse was required to leach the same amount from a sandy soil (Sharom et al. 1980). The positive contribution of SOM to carbaryl sorption is evident in Table 3; the sorption capacity ($K_f$) was reported to increase with SOM content in Indian soils (Jana and Das 1997).

A comparison of carbaryl sorption to soils from four countries is presented in Table 4. Although organic carbon influences carbaryl sorption (i.e., $K_d$), a positive correlation was not observed by Ahmad et al. (2001a). However, in a later study Ahmad et al. (2001b) reported a positive, highly significant, correlation of organic carbon-normalized sorption capacity ($K_{oc}$) with the aromatic content of SOM. $K_d$ values similar to those presented in Table 4 have been reported (Bondarenko and Gan 2004), indicating the sorption of carbaryl to soils is not very important.

Carbaryl sorption has been predicted to be highly reversible because, in contrast to chemisorption, it is proposed to be nonspecific (Rajagopal et al. 1984). This property, along with reported low $K_d$ values, indicates that soils do not possess the potential to significantly retard carbaryl movement, over time, into either surface

**Table 3** The relationship between soil organic matter (SOM) and the sorption capacity ($K_f$) in four different soils from India

| Soil | SOM (%) | $K_f$ (µg/g)/( µg/mL) |
|---|---|---|
| Ultisol 1 | 0.40 | 0.308 |
| Inceptisol 2 | 1.10 | 1.916 |
| Ultisol 2 | 1.16 | 2.175 |
| Inceptisol 1 | 1.70 | 2.490 |

*Source*: Jana and Das (1997)

**Table 4** Distribution coefficients ($K_d$) for carbaryl in several soils

| Soil | Organic carbon (g/kg) | $K_d$ | Sand:silt:clay (%) |
|---|---|---|---|
| Pakistan 2[a] | 2.79 | 0.99 | 22:60:18 |
| Australian 2[a] | 3.0 | 0.19 | 92:5:3 |
| United Kingdom 2[a] | 8.9 | 1.09 | 10:67:23 |
| Pakistan 1[a] | 13.82 | 59.67 | 22:51:27 |
| Australian 1[a] | 58 | 23.02 | 63:16:21 |
| United Kingdom 1[a] | 83.8 | 8.80 | 18:39:43 |
| California 1[b] | – | 43.4 | – |
| California 2[b] | – | 47.7 | – |

*Sources*: [a] Ahmad et al. (2001a); [b] California 1 and 2 represent sediment from San Diego Creek and Bonita Creek in California, USA (Bondarenko and Gan 2004).

waters or groundwater; other fate processes (i.e., abiotic or biotic degradation) play an important role in its dissipation.

# 4 Degradation

## 4.1 Abiotic

**Hydrolysis**

Carbaryl is effectively hydrolyzed in water, undergoing a 50% loss at 20°C (pH = 8) in 4 d (Rajagopal et al. 1984). Earlier studies have reported similar degradation times: 6 d in flowing canal water (Osman and Belal 1980), and 1 wk in river water (Eichelberger and Lichtenberg 1971). These investigators and others (Ghauch et al. 2001) have also shown that hydrolysis of the insecticide increases with elevated temperature. Hydrolytic degradation is initiated by hydroxyl radical attack (Fig. 2; Wang and Lemley 2002), producing 1-naphthol as the primary degradation product (Osman and Belal 1980).

**Photolysis**

Carbaryl has been reported to be photolyzed into 1,2-naphthoquione, 1,4-naphthoquinone, 2-hydroxy-1,4-naphthoquinone, and 7-hydroxy-1,4-naphthoquinone (Fig. 3; Brahmia and Richard 2003); conversion to 1-naphthol via hydroxyl radical attack has also been observed in organic solvents (e.g., acetonitrile and methanol; Fig. 3F). In water, carbaryl produces naphthoxyl radicals, which confirm the cleavage of the carbon–oxygen bonds. However, in oxygen-rich water, solvated electrons could be

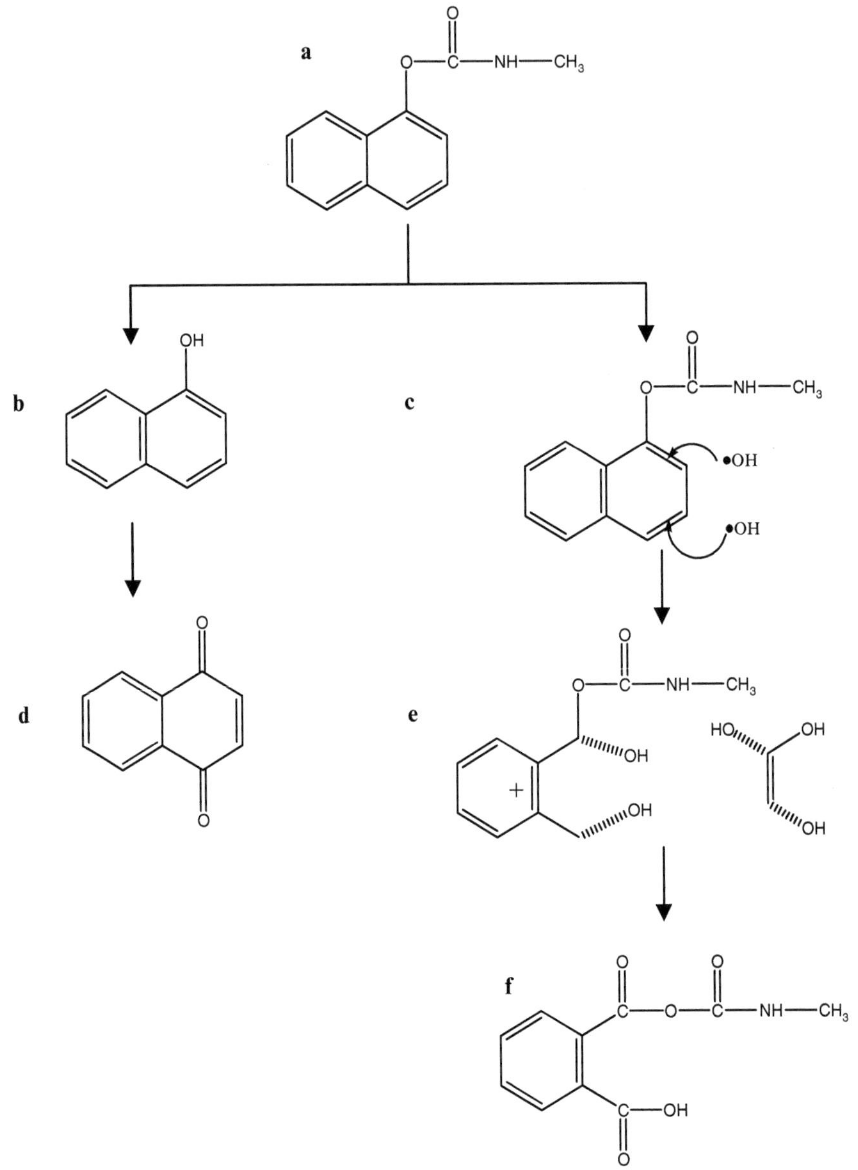

**Fig. 2** The degradation pathways (Wang and Lemley 2002) of carbaryl (**a**) by hydroxyl radical attack (**c, e**) showing the degradation products: 1-naphthol (**b**), 1,4-naphthoquinone (**d**), and phthalic acid-O-yl N-methylcarbamate (**f**)

transformed into superoxide anions that can recombine with radical cations or with 1-naphthoxyl radicals. Both reactions are expected to produce naphthoquinones after reduction (Brahmia and Richard 2003).

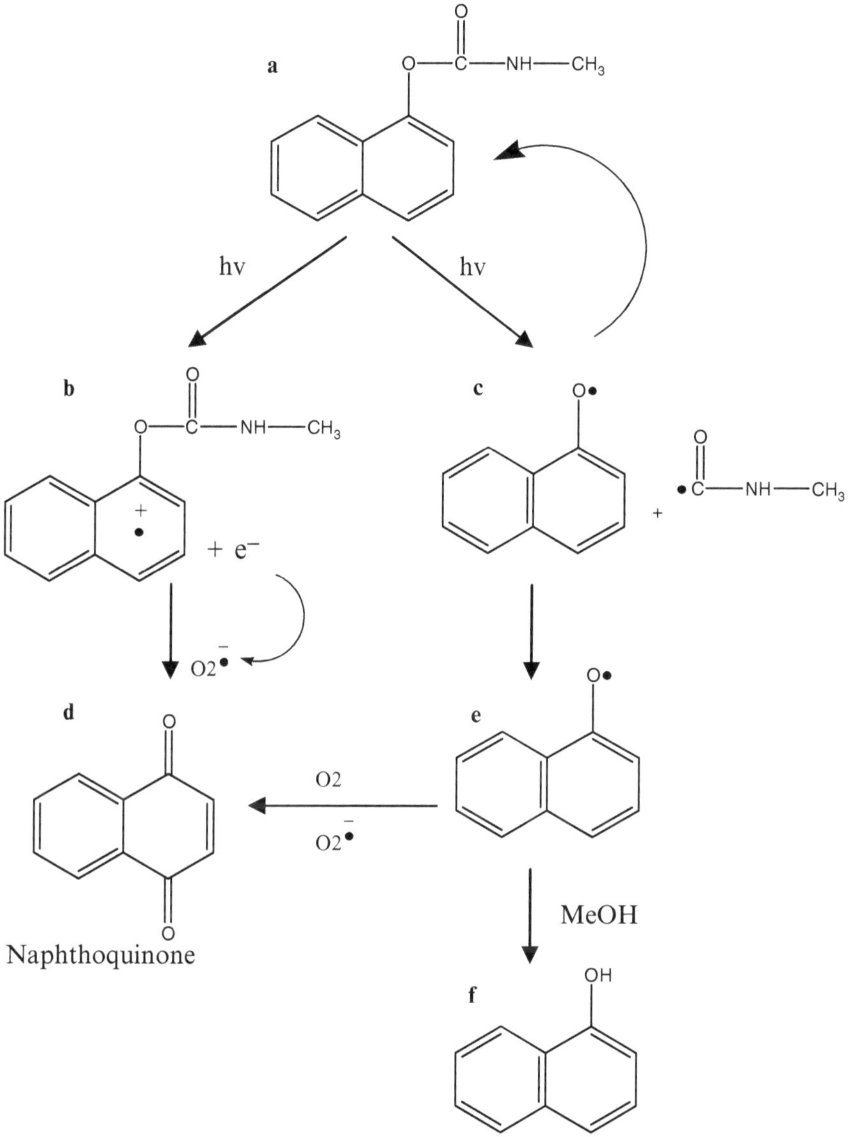

**Fig. 3** Proposed (Brahmia and Richard 2003) photolytic degradation pathway for carbaryl (**a**). The parent compound is distributed into radicals (**b, c**) via photolytic processes. 1-Naphthoxyl (**c**) may then react with oxygen to yield naphthoquionone (**d**) or 1-naphthol (**f**)

Indirect photolysis of carbaryl has been reported by Miller and Chin (2002). They found that photo-enhanced degradation was both seasonally and spatially dependent. Nitrate and dissolved organic matter (DOM) were primary constituents responsible for the formation and reaction of hydroxyl radicals with carbaryl (Miller and Chin 2002).

## 4.2 Biotic

**Microbial**

The microbial degradation of carbaryl has been reported in several studies. For instance, ring $^{14}$C-labeled carbaryl degraded at a constant rate in 120 d, leaving behind 15% –20% of the parent compound in soil as monitored by the release of $^{14}CO_2$ (Rodriguez and Dorough 1977). Shorter degradation times have been observed by Menon and Gopal (2003) in that carbaryl was found to dissipate to below detection levels within 45 d ($DT_{50}$ = 14.93). However, this relatively rapid degradation was attributed to high temperatures and precipitation. Still shorter $DT_{50}$s have been reported, ranging from 0.15 d (Wolfe et al. 1978) to several days (Tomlin 2003). Under aerobic soil conditions, reported $DT_{50}$s were 7–14 d in sandy loam and 14–28 d in clay loam (Tomlin 2003). Bondarenko and Gan (2004) observed aerobic $DT_{50}$ values of 1.8 and 4.9 d in soils containing organic matter at 1.8% (sand:silt:clay = 76:15:9) and 1.25% (sand:silt:clay = 46:32:22), respectively. Pseudo-first-order kinetics have been applied to describe the microbial degradation of carbaryl in moist soils (Venkateswarlu et al. 1980). However, inhibition of its degradation can occur when ammonium is added to the enrichment cultures (Rajagopal et al. 1983), possibly indicating that carbaryl may serve as a source of essential nitrogen for microbes.

Degradation has been observed to be more rapid in flooded (anaerobic) soils than aerobic soils; the $DT_{50}$ was reported to be 13–14 d in flooded soils and 23–28 d in aerobic soils (Venkateswarlu et al. 1980). Rajagopal et al. (1983) observed a $DT_{50}$ of 10–15 d in both submerged laterite and sodic soils, and that degradation was faster in soils previously treated with carbaryl. However, recently Bondarenko and Gan (2004) reported that under anaerobic conditions carbaryl was slowly degraded, with $DT_{50}$ values from 125 to 746 d, depending on soil conditions, sorption capacity, and aging of the soil with the insecticide.

Mechanisms of degradation have also been reported. Karinen et al. (1967) showed carbaryl ring degradation through 1-naphthol, its primary degradate, to $CO_2$. Thus, ring hydroxylation is the first step in microbial degradation. Such findings are supported by Rajagopal et al. (1983), who noted that hydrolysis was the major pathway of degradation in flooded (anaerobic) soils (see Fig. 4). The primary product, 1-naphthol, has a $DT_{50}$ of approximately 12–14 d (Menon and Gopal 2003), and can be further transformed to phenolic radicals, which polymerize to organic matter in soils (Rajagopal et al. 1984). Complete degradation from carbaryl to maleylpyruvate is reported for an isolated *Micrococcus* species (Fig. 4) by Doddamani and Ninnekar (2001).

Other microbial strains capable of degrading carbaryl have also been identified, including bacteria of the genera *Achromobacter, Pseudomonas (e.g., P. cepacia), Arthrobacter*, and *Xanthomonas* (Venkateswarlu et al. 1980; Rajagopal et al. 1984). Degradation by the fungus *Penicillium implicatum* has also been demonstrated (Menon and Gopal 2003). However, carbaryl has been found to inhibit the growth of some strains of rhizobia (Rajagopal et al. 1984).

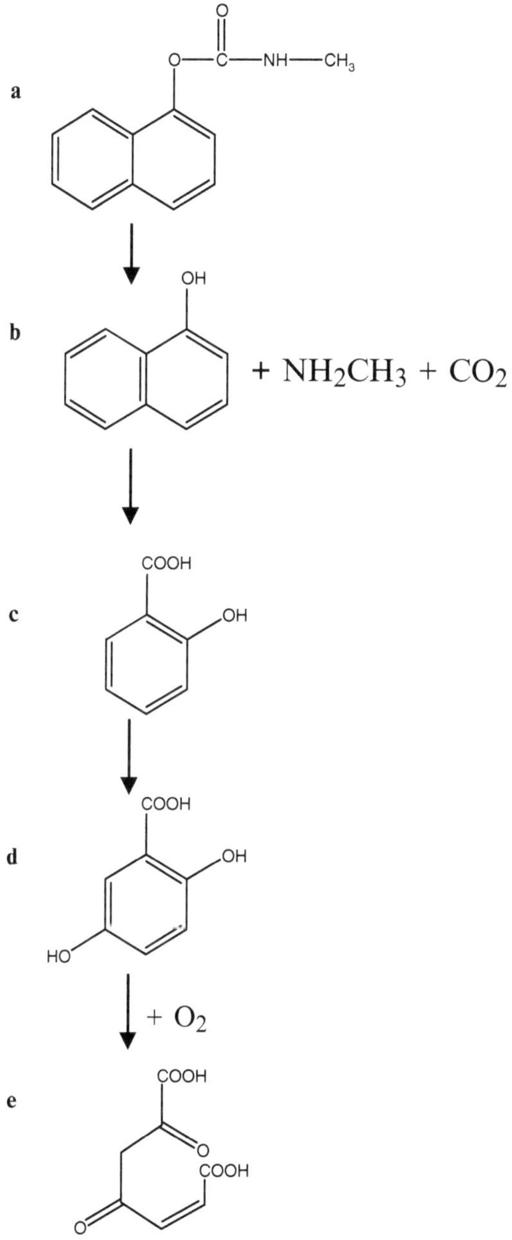

**Fig. 4** Proposed degradation pathway of carbaryl by Micrococus sp. (Doddamani and Ninnekar 2001). Carbaryl (**a**) is reduced to 1-naphthol and methylamine (**b**), which is then degraded to salicylic (**c**) and gentrisic (**d**) acid. The acids are then oxidized to maleylpyruvate (**e**)

## Higher-Order Organisms

The metabolism of carbaryl has been extensively studied in mammals. In general, it does not accumulate in mammalian tissue and is rapidly metabolized to less-toxic substances, particularly 1-naphthol, which are eliminated in urine and feces (Tomlin 2000). The main pathways include oxidation, via hydroxylation and epoxidation, and hydrolysis (Carpenter et al. 1961; Dorough and Casida 1964). For instance, hydrolysis of carbaryl by earthworms forms 1-naphthol (Stenersen 1992). A hydrolytic mechanism has been proposed by Sogorb et al. (2002) in which carbaryl reacts with tyrosine residues on rabbit serum albumin to yield 1-naphthol and carbamylated albumin. Water molecules then attack the carbamylated complex, releasing carbamic acid and free enzymes, the latter of which are involved in a new catalytic cycle; carbamic acid probably decomposes to $CO_2$ and methylamine (Sogorb et al. 2002). Metabolites detected in urine of human workers exposed to carbaryl were both 1-naphthyl-glucoronide and 1-naphthylsulfate (Sogorb et al. 2004). Carbaryl metabolism in human liver microsomes and by cytochrome P450 isoforms was investigated by Tang et al. (2002). They found three major metabolites: 5-hydroxycarbaryl, 4-hydroxycarbaryl, and carbaryl methylol (Fig. 5). Interestingly, these are the same as those formed by plants (Tomlin 2003).

Factors inhibiting enzymatic hydrolysis have also been noted. For instance, Sogorb et al. (2004) suggest that long-chain fatty acids are better inhibitors of carbaryl hydrolysis than shorter ones. Several organic compounds can inhibit its hydrolysis as well. For example, the organophosphorus insecticide chlorpyrifos inhibits carbaryl

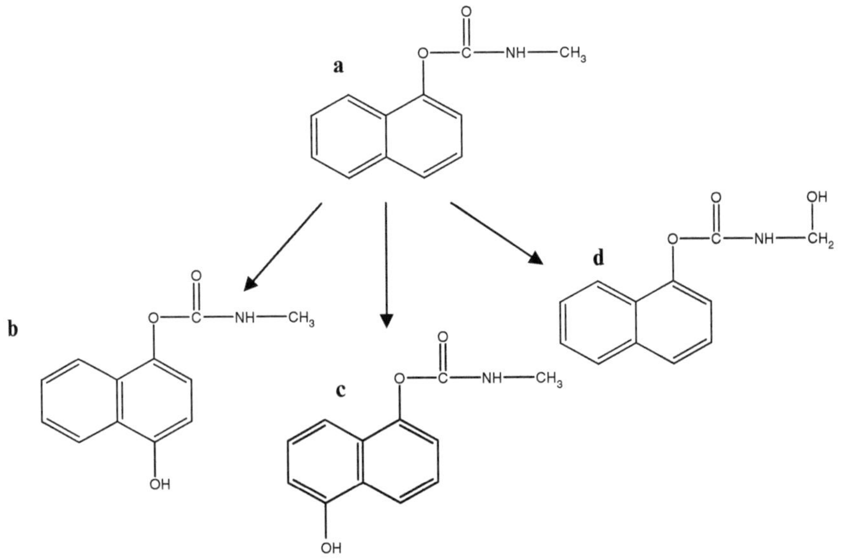

**Fig. 5** The cytochrome P450-dependent metabolism (Tang et al. 2002) of carbaryl (**a**) to 4-hydroxycarbaryl (**b**), 5-hydroxycarbaryl (**c**), and carbaryl methylol (**d**)

hydrolysis (Tang et al. 2002); the activated form of ethyl parathion, paraoxon, inhibits hydrolysis by 44% (Sogorb et al. 2004).

Carbaryl has been observed to react with certain nitrogen-containing compounds, such as sodium nitrite, to form nitrosocarbaryl, which has been found to cause skin cancer when painted on mice (Deutsch-Wenzel et al. 1985), and cancer of the stomach in rats (Lijinsky and Schmahl 1978; Lijinsky and Taylor 1976). Nitrosocarbaryl belongs to the $N$-nitrosoamines class of chemicals, of which some 70% to date have been found to be carcinogenic (Cox 1993).

# 5 Toxicity

## 5.1 Insects and Aquatic Organisms

Carbaryl is highly effective for controlling insect pests. For example, it is used to control several mammalian ectoparasites, including the cattle tick *Boophilus microplus*. The tick is endemic to Mexico, having been eradicated from the U.S. in 1961 (Li et al. 2005). Several strains of *B. microplus* are highly susceptible to carbaryl; $LC_{50}$s range from 0.0025% to 0.0031% (Li et al. 2005). Carbaryl is also highly toxic to the honeybee, with a topical $LD_{50}$ of 1 μg (Tomlin 2003).

Although carbamate pesticides do not persist in the environment, there may still be short-term cumulative effects on the reproduction of aquatic organisms. For instance, Tripathi and Singh (2004) found that doses of 2, 5, and 8 mg/L altered biochemical function in the nervous, hepatopancreatic, and ovotesticular tissues of the snail *Lymnaea acuminate*. Specifically, glycogen, pyruvate, total protein, and nucleic acid levels were reduced after 96 hr exposure, whereas lactate and free amino acid levels were increased (Tripathi and Singh 2004). Carbaryl can also affect embryo development. For example, Tripathi and Singh (2004) reported that the number of eggs produced by the freshwater snail *Lymnaea acuminate* was reduced by 49% at 2 mg/L; no eggs were laid at 5 or 8 mg/L. The rate of neonatal survival was also significantly reduced by 53% after exposure of hatchlings for 28 d to 2 mg/L. In a similar study, Todd and Van Leeuwan (2002) found that the average mortality of zebrafish eggs (*Danio rerio*) was reduced (~20%) after low-level exposures (<0.05 mg/L). Although the insecticide did not directly kill embryos, it had a significant effect on embryo size.

When zebrafish were exposed to 0.017 mg/L carbaryl, they developed more slowly and hatched later compared to the controls; delayed hatching exposes the embryos to predation. The toxicity of carbaryl to several aquatic species is summarized in Table 5. Note that carbaryl is toxic to the water flea, shrimp, and freshwater snail at parts per billion (ppb) levels and to fish at parts per million (ppm) levels. The results suggest that this insecticide should not be used in or near water bodies, particularly during the rainy season.

**Table 5** The aquatic animal toxicology of carbaryl

| Aquatic organism | Test | Concentration (mg/L unless noted) |
|---|---|---|
| Juvenile trout[a] | 96-hr $LC_{50}$ | 4.27–6.18 |
| Toad larvae[a] | 96-hr $LC_{50}$ | 17.68–34.77 |
| Juvenile trout[a] | $IC_{50}$ | 19 µg/L |
| Toad larvae[a] | $IC_{50}$ | 7.580 |
| Rainbow trout[b] | 96-hr $LC_{50}$ | 1.3 |
| Sheepshead minnow[b] | 96-hr $LC_{50}$ | 2.2 |
| Bluegill sunfish[b] | 96-hr $LC_{50}$ | 10 |
| Mysid shrimp[b] | 96-hr $LC_{50}$ | 5.7 µg/L |
| Eastern oyster[b] | 48-hr $LC_{50}$ | 2.7 |
| Shrimp larvae[c] | 96-hr $LC_{50}$ | 30 µg/L |
| Common carp[d] | 96-hr $LC_{50}$ | 7.85 |
| Freshwater snail[e] | 24-hr $LC_{50}$ | 20.05 |
| Freshwater snail[e] | 96-hr $LC_{50}$ | 14.19 |
| Water flea (*Bosmina longirostris*)[f] | 24-hr $LC_{50}$ | 8.6 µg/L |
| Water flea (*Bosmina fatalis*)[f] | 24-hr $LC_{50}$ | 4.1 µg/L |
| Water flea predator (*Leptodora kindtii*)[f] | 24-hr $LC_{50}$ | 3.6 µg/L |

*Sources*: [a] Ferrari et al. (2004); [b] Tomlin (2003); [c] Reyes et al. (2002); [d] De Mel and Pathiratne (2005); [e] Tripathi and Singh (2001); [f] Sakamoto et al. (2005).

## 5.2 Animals

Acute oral $LD_{50}$s as well as the irritation and sensitization properties for carbaryl are presented in Table 6. The lower $LD_{50}$s in rats reported for intraperitoneal (IP) versus oral exposure suggest that either hepatic (or possibly gastrointestinal) metabolism and excretion mediate the response to carbaryl, or that absorption from the IP route is faster than by the oral route, resulting in temporally higher blood and tissue concentrations.

The most detailed accounts of rodent responses to low, orally gavaged doses of carbaryl come from a series of studies with Sprague–Dawley rats reported by a single laboratory in the 1990s. Definitive acute effects, including pinpoint pupils, ataxic gait, tremors, a reduction in motor activity counts, and body weight gain decrements were noted at doses as low as 10 mg/kg; doses up to 125 mg/kg produced salivation and/or wet muzzle, overall gait incapacity, and an mpaired visual placing response (Brooks et al. 1995; Robinson and Broxup 1997). In addition, there were increases in hindlimb splay (males; at high doses), arousal and number of rears, positional passivity, auricular startle responses, and males lying on their ventral surface. Finally, there were decreases in locomotor activity, extensor thrust, tail- and toe-pinch responses, urination and defecation in males, vocalization upon cage removal in females, forelimb and hindlimb grip strength, and body temperature. These effects largely abated by day 7 and 14. Beyrouty (1992) reported similar effects in orally gavaged Sprague–Dawley rats, although generally not at doses lower than 40 mg/kg.

**Table 6** The acute toxicity and primary irritation properties of technical grade carbaryl

| Animal | Test | US EPA toxicity category | Amount (mg/kg unless noted) |
|---|---|---|---|
| Rat (male)[1-5] | Oral $LD_{50}$ | II | 233–840 |
| Rat (Female)[1-5] | Oral $LD_{50}$ | II | 246–610 |
| Mouse[4] | Oral $LD_{50}$ | II | 108–650 |
| Rabbit[1] | Oral $LD_{50}$ | III | 710 |
| Guinea pig[1,4] | Oral $LD_{50}$ | II | 280 |
| Dog[4] | Oral $LD_{50}$ | II | 250–795 |
| Cat[4] | Oral $LD_{50}$ | II | 125–250 |
| Swine[4] | Oral $LD_{50}$ | III | 1,500–2,000 |
| Deer[4] | Oral $LD_{50}$ | II | 200–400 |
| Monkey[4] | Oral $LD_{50}$ | III | >1,000 |
| Rat[4] | Dermal $LD_{50}$ | III | >2,000 to >5,000 |
| Rabbit[2,6] | Dermal $LD_{50}$ | III | >2,000 |
| Rat[7] | 4-hr Inhalation $LC_{50}$ | III | 0.873 mg/L |
| Rat[8] | 4-hr Inhalation $LC_{50}$ | III | 2.50 mg/L |
| Rat (female)[9] | Oral $LD_{50}$ | II | 437.5 |
| Mouse (female)[9] | Oral $LD_{50}$ | III | 515 |
| Guinea pig[10,11] | Dermal sensitization | Negative | – |
| Rabbit | Dermal irritation | IV | – |
| Rabbit | Eye irritation | IV | – |
| Rat (male adult)[12] | Intraperitoneal $LD_{50}$ | – | 64 |
| Rat (male weanling at 23 d)[12] | Intraperitoneal $LD_{50}$ | – | 48 |

*Sources*: [1]Mellon Inst. (1957); [2]Union Carbide (1983a–d); [3]Union Carbide (1985); [4]Cranmer (1986); [5]Larson (1987b); [6]Larson (1987a); [7]Holbert (1989); [8]Dudek (1985); [9]Rybakova (1966); [10]Larson (1987c); [11]US EPA (2002); [12]Brodeur and DuBois (1963).

An attempt to determine the time of peak brain ChE inhibition after oral gavage (at 10 mg/kg) showed that enzyme activity was suppressed by 46% of control levels within 0.5 hr in males and by 54% in females (Brooks and Broxup 1995). The degree of inhibition declined steadily after 1 hr, although inhibition was still evident at the high dose of 125 mg/kg at 24 hr. Except for one 10 mg/kg male exhibiting muzzle/urogenital staining at 0.5 hr, behavioral and/or clinical signs, including tremors and autonomic changes, were seen only at 50 and 125 mg/kg. The time to peak effect was also determined to be in the 0.5–1 hr range, generally lessening after that time.

Desi et al. (1974) noted a biphasic response during 50 d dietary exposure of Wistar rats to carbaryl at 10 or 20 mg/kg/d. During the first 15 d, performance times in T-mazes actually improved (i.e., the learning times decreased), whereas after that point performance then worsened (i.e., learning times increased). The authors ascribed the initial improvements to "enhanced irritability" of the central nervous system (CNS). Their conviction that the CNS was the main site of action was strengthened by the observation that "the animals were able to move quickly even during the second period" (i.e., the period of decreased maze function). This observation was further supported by the evidence that electroencephalograms and brain cholinesterase activities were altered by carbaryl.

Conversely, Austin (2002) was unable to elicit behavioral or morphological signs in Sprague–Dawley rats during 4 wk of daily (6–7 hr/d) dermal treatment with up to 100 mg/kg/d carbaryl, although that dose resulted in a body weight gain decrement during the day 5 to day 12 period. Weekly postdose measurements of erythrocyte ChE activities revealed significant suppression on days 5 and 12 at 50 and 100 mg/kg/d; however, by 26 d no such effects were evident. Brain ChE activities measured on day 26 revealed up to 15% inhibition at 50 mg/kg/d and 24% inhibition at 100 mg/kg/d. However, it was not clear if these effects were acute or if they occurred after several daily doses.

Dogs dietarily exposed to doses as high as 35 mg/kg/d carbaryl for 1 yr did not exhibit clinical signs (Hamada 1987). Nonetheless, ChE activities were suppressed at all time points, often by significant margins. For brain ChE, which was measured only at the 52-wk study termination point, the level of inhibition reached 36% at the high dose, although a significant 20% level of inhibition was noted in females even at the low dose of 3.7 mg/kg/d. Erythrocyte ChE inhibition up to 56% was noted at the high dose (week 5), with nonstatistically significant inhibition of up to 14% (week 13) noted at the low dose. Plasma cholinesterase inhibition reached 66% (week 5) at the high dose, with significant inhibition of up to 23% (week 13) noted at the low dose.

Carbaryl administered to mice through the diet over a 2-yr period induced significantly elevated hemangiosarcomas and hemangiomas in males at the mid and high doses of 145 and 1,249 mg/kg/d (Hamada 1993a). Females experienced a significant increase in tumors at the high dose only; a nonstatistically significant increase was noted in males at the low dose of 14 mg/kg/d. In addition, significant elevations of hepatocellular carcinomas and adenomas were detected in high-dose females, as were kidney tubular adenomas and carcinomas in high-dose males. Other observations included unscheduled deaths (females at the high dose), clinical signs (at both mid and high doses), ChE suppression (at both mid and high doses, with a suggestion of a low-dose effect), and body weight effects (at the high dose). Nononcogenic histopathology was noted in the bladder (intracytoplasmic droplets/pigment; at mid and high doses), eye (cataracts; at the high dose), and spleen (abnormal pigmentation; at the high dose). Although there is some question about the biological significance of the effects at the high dose, which may have exceeded the maximum tolerated dose, the presence of hemangiosarcomas and hemangiomas at the mid and low doses demonstrated the carbaryl carcinogenic potential.

Dietary exposure of rats at doses up to 485 mg/kg/d for 2 yr led to hyperplastic and neoplastic lesions in the urinary bladder of both sexes, particularly at the high dose (Hamada 1993b). These effects included hyperplasia, transitional cell papillomas, transitional cell carcinomas, squamous metaplasia, high mitotic index, and atypia.

Carbaryl tested positive in one of five gene mutation studies (Lawlor 1989; Grover et al. 1989; Young 1989; Ahmed et al. 1977a; Onfelt and Klasterska 1984), four of six chromosomal aberration studies (Weil 1972; Murli 1989; Ishidate and Odashima 1977; Marshall 1996; Grover et al. 1989; Soderpalm-Berndes and Onfelt 1988), and two of four DNA damage studies (Ahmed et al. 1977b; Cifone 1989;

Onfelt and Klasterska 1984; Sagelsdorff 1994). The insecticide should thus be viewed as potentially genotoxic. Because virtually all the positive studies were performed *in vitro*, they were considered less relevant than the *in vivo* studies on whole organisms. One study in V79 Chinese hamster fibroblasts showed that, as is carbaryl, 1-naphthol was toxic and induced c-mitosis, an aberrant form of mitosis that may reflect effects on mitotic spindle formation (Soderpalm-Berndes and Onfelt 1988).

Nitrosocarbaryl, the potentially carcinogenic derivative, was formed more readily from carbaryl in the very acidic guinea pig stomach versus the less acidic rat stomach (Rickard and Dorough 1984). In addition, Eisenbrand et al. (1975) demonstrated that nitrosocarbaryl could be produced from carbaryl and nitrite under acidic *in vitro* conditions. Nitrosocarbaryl has been reported to cause single-strand breaks in cultured human fibroblasts (Regan et al. 1976).

Two reproductive toxicity studies by Pant et al. (1995, 1996) used Wistar rats exposed to carbaryl by daily gavage at 50 mg/kg/d for 90 d (5 d/wk). Impacts noted on testicular enzymes, sperm counts, sperm motility, sperm morphology, and testicular morphology supported the conclusions of two earlier studies by Rybakova (1966) and Shtenberg and Rybakova (1968), as well as the suggestive epidemiological results summarized in the following section. Narotsky and Kavlock 1995) observed fetal resorption in 2 of 13 pregnant rats and a 6% weight decrement in pups exposed *in utero* to carbaryl by gavage at 104 mg/kg/d, a dose that also caused overt toxicity in the dams. A study in gerbils demonstrated impairments in several reproductive indices at and above a dietary dose of about 160 mg/kg/d (Collins et al. 1971). However, Tyl et al. (2001) did not observe effects on reproductive indices in rats subjected to dietary carbaryl, although there was some suggestion that pup survival through 14 d was reduced at the mid and high doses of 21–36 mg/kg/d and 92–136 mg/kg/d, respectively. No attempt to ascertain sperm morphology was undertaken in the Tyl el al. (2001) study.

Aside from developmental delays, possibly mediated by suppressed maternal weight gains, the studies of Repetto-Larsay (1998) and Tyl et al. (1999) provided minimal evidence for carbaryl-induced developmental toxicity in rats and rabbits, although omphalocele was present at gavage doses 150 and 200 mg/kg/d in an older rabbit study (Murray et al. 1979). However, Smalley et al. (1968) demonstrated severe maternal and fetal effects in beagle dogs following dietary exposure to carbaryl during gestation, including (1) increased dystocia at all dose levels (3.125–50 mg/kg/d); (2) three mothers sustaining total fetal deaths (one each at 6.25, 25, and 50 mg/kg/d); (3) decreased pup weight gains in the combined treatment groups; (4) decreased conception rate at the high dose; (5) no pups born alive at the high dose; (6) decreased percentage of pups weaned (effect possibly present at levels as low as 6.25 mg/kg/d), and (7) increased litters with pups bearing abnormalities (6.25 mg/kg/d and above). Observed abnormalities included abdominal-thoracic fissures with varying degrees of intestinal agenesis and displacement, varying degrees of brachygnathia, ecaudate pups (i.e., without a tail), failure of skeletal formation, failure of liver development, and superfluous phalanges.

## 5.3 Humans

Baron (1991) reviewed several studies involving systemic carbaryl exposures in humans. No effects were observed in one acute oral study with males at doses as high as 2 mg/kg. In another study, a scientist investigating possible antihelmintic properties of carbaryl ingested approximately 2.8 mg/kg. Epigastric pain followed by profuse sweating began after 20 min, followed by lassitude and vomiting. Recovery was evident after 1 hr (although 3 mg antidotal atropine had been ingested by then) and completed by 2 hr. Similarly, a researcher who intentionally ingested carbaryl at 5.45 mg/kg on an empty stomach experienced vision changes, nausea, and lightheadedness within 80–90 min post dose. Despite two doses of atropine, profuse sweating, hyperperistalsis, and weakness were present by 97 min, with maximal symptomology at 2 hr; complete reversal of symptoms occurred by 4 hr.

Dickoff et al. (1987) reported responses in a patient found comatose 3 hr after ingestion of approximately 500 mg/kg carbaryl. The following overt symptoms were noted within 1–2 d: salivation, miosis, eyelid twitching/fasciculation/abnormal movements, flaccid tone, pulmonary edema, diarrhea, incontinence, low systolic blood pressure and body temperature, elevated heart rate, intubation for control of breathing and bronchial secretion, lack of responsiveness to voice or pain, lack of spontaneous limb movement, ankle clonus (but no plantar response), diarrhea, abdominal cramping, and dark brown heme-negative urine. Within 3–6 d, the following were apparent: prickling foot/leg/hand pain and other diffuse pain, leg paralysis, absent tendon reflexes, occasional rapid involuntary flexion of knees and hips, hand weakness, inability to sit alone, sensory loss in extremities, pseudoathetotic arm movements, proximal right leg movements, no voluntary motor units or only single unit recruitment patterns in distal leg muscles, no abductor digiti quinti response decrement after repetitive ulnar nerve stimulation, and symmetrically diffuse electroretinogram. After 3 wk, responses included impaired finger strength, inability to stand, plantar responses were flexor, persistent tenderness to distal palpation, marked impairment to pin and vibration below the knees, and absent position sense in toes and impaired in ankles (normal in fingers); after 5 wk responses consisted of bilateral foot drop, no volitional motor units below knees, pin sensation absent in lower legs, toe position/vibration absent, diminished compound muscle action potential amplitudes in tested nerves, slight slowing of leg conduction velocities, low amplitude evoked sensory nerve responses, increased insertional activity in electromyogram, muscle fibrillations and positive waves, and periods of diffuse/symmetric slowing with electroencephalograms. By 9 mon the patient experienced a return to normal strength except for bilateral ankle/toe weakness; jerk responses were elicited in triceps only. There was a persistent loss of toe vibration/proprioception, pin and touch responses were reduced to midcalf, and electroencephalograms exhibited normal characteristics. The authors suspected that carbaryl induced a delayed polyneuropathy similar to the delayed syndrome known to occur with organophosphate exposures.

Branch and Jacqz (1986) described the extensive toxic sequelae in a 75-yr-old man exposed accidentally, but over a prolonged period, to a 10% carbaryl dust formulation that occurred during and after six monthly treatments of his house to combat fleas; of particular concern was the evidence for permanent neurological damage. Tomography undertaken several years later revealed progressive dilation of the cerebral ventricles "associated with a reduction in cerebral function and intellectual capacity." Interestingly, the patient's wife and son experienced some initial symptoms, although they resolved without the appearance of longer-term disabilities. Continuing treatment of the patient with cimetidine to ameliorate gastric symptoms was a possible confounding factor.

Wyrobek et al. (1981) conducted an epidemiological investigation of testicular function among carbaryl-exposed factory workers. This study failed to establish a clear connection between exposure and seminal defects, although the data suggested an increase in oligospermia (defined as a sperm count $<20 \times 10^6$/mL) and teratospermia (defined as exhibiting >60% abnormal sperm forms). A more recent study of factory workers from China demonstrated significantly higher levels of sperm chromosomal aberrations and DNA damage in an occupationally exposed population (Xia et al. 2005). Meeker et al. (2004a,b) noted an association between 1-naphthol levels in the urine and a series of sperm toxicity parameters, including decreased sperm concentrations, decreased sperm motility, and increased DNA single-strand breaks resulting in high tail % (a measure of the proportion of DNA in the electrophoretic tail) in comet assays. However, it was not known if the 1-naphthol originated as a metabolite of carbaryl or naphthalene or had another source. The possible reproductive effects of carbaryl were considered in an epidemiological study of pregnancy outcomes following exposure to males from farm families in Ontario, Canada (Savitz et al. 1997). In conjunction with carbaryl exposure the adjusted odds ratio for miscarriage increased, suggesting that exposure of reproductive-aged males could result in clinically manifested reproductive impacts.

Several potential carbaryl exposure scenarios were considered in the U.S. Environmental Protection Agency's recent interim health hazard assessment (US EPA 2004). In addition, the USDA Pesticide Data Program documented the presence of both carbaryl and 1-naphthol in many raw agricultural commodities destined for sale within the United States (http://www.ams.usda.gov/pdp), and these are highlighted in the US EPA recent dietary risk assessment on carbaryl (US EPA 2003), as well as in the California Department of Pesticide Regulation (CDPR) upcoming dietary risk assessment, which assesses dietary exposure and the possibility of toxic responses to California residents. From 1992 to 2005, the California Pesticide Illness Surveillance Program documented 32 illness incidents with a reasonable possibility of association with carbaryl exposure alone, as well as 57 others with a reasonable possibility of association with carbaryl in combination with other pesticides (http://www.cdpr.ca.gov/docs/whs/pisp.htm).

# 6 Mammalian Toxicokinetics and Metabolism

Struble (1994) studied the disposition of radiolabeled carbaryl in Sprague–Dawley rats following administration by oral gavage and reported that it was excreted primarily via urine during the first 24 hr, although substantial residues appeared in feces and exhaled air as $CO_2$ (detectable when the label resided on the carbonyl or N-methyl carbon, but not when on the naphthalene ring). Metabolites were conjugated with sulfate or glucuronic acid. For animals receiving 1 mg/kg, about half the dose was detected in the urine during the first 6 hr, 80% –90% by 24 hr, and only slightly more by 168 hr. For animals receiving 50 mg/kg, urinary excretion was somewhat slower: 12% –20% by 6 hr, 64% –69% by 24 hr, and 78% –81% by 168 hr. Fecal excretion was also significant: by 168 hr, some 6% –13% of the dose appeared in the feces.

Krolski et al. (2003) examined the kinetics of [naphthyl-1-$^{14}$C]-carbaryl in blood and other tissues after oral (1.08 or 8.45 mg/kg), dermal (17.25 or 102.95 mg/kg) and intravenous (i.v.; 0.80 or 9.20 mg/kg) exposure. Peak levels of radioactivity were detected in blood at 15 and 30 min for both the low and high dose oral treatments, respectively; at 4 and 12 hr for dermal application; and were already maximal by the first time point (5 min) after i.v. injection. By 24 hr after oral dosing, radioactivity levels had decreased to 0.81% –2.4% of their peaks in blood fractions (both doses), 0.60% –2.4% in brain (both doses), 0.67% in liver (high dose only), and 0.32% in fat (high dose only). With dermal dosing, radioactivity levels had decreased to 15.9% –25.8% of their peaks in blood fractions (both doses), 27.1% – 30.6% in brain (both doses), 24.4% in liver (high dose only), and 15.6% in fat (high dose only) by 24 hr. Finally, with an i.v. dose, by 24 hr radioactivity levels had decreased to 4.6% –10.5% in blood fractions (both doses), 1.1% –1.3% in brain (both doses), 5.7% in liver (high dose only), and 0.72% in fat (high dose only).

The pharmacokinetic disposition of carbaryl in mice (Totis 1997; Valles 1999), guinea pigs (Knaak et al. 1965), and sheep (Knaak et al. 1968) appeared generally similar to those in the rat, although there were significant technical problems with the guinea pig and sheep studies, as they employed few animals and left large fractions of the administered dose unanalyzed. A possible exception to the rat model was the dog, where approximately equal fractions of an oral carbaryl dose were excreted after 24 hr in urine and feces (Knaak and Sullivan 1967). However, these data, reported by the same investigators, suffered from similar problems. Speculation about the tendency toward tumor formation at high doses in a more recent mouse study (see above) centered on a shift in the urinary metabolite pattern at the comparatively high dose of 8,000 ppm (~1600 mg/kg/d), where there were increases in metabolites derived from epoxide intermediates (Valles 1999).

Three major metabolic pathways, presumably hepatic, were identified in the rat (Struble 1994): (1) arene oxide-mediated hydroxylation and subsequent conjugation; (2) hydrolytic decarbamylation, to form 1-naphthol, and subsequent conjugation; and (3) oxidation of the N-methyl group. Three urinary metabolites found in rat urine [1-naphthyl glucuronide, 1-naphthyl sulfate, and 4-(methyl-carbamyloxy)-1-naphthyl glucuronide] were not found in dog urine (Knaak and Sullivan 1967). In addition, few hydrolytic products were found in the urine of a singly dosed monkey (Knaak et al. 1968). The toxicological significance of these species differences is not clear. Humans appear to have the ability to decarbamylate carbaryl, as factory workers were found to excrete 1-naphthyl glucuronide and 1-naphthyl sulfate. This finding led to speculation that humans are similar to rats in their pharmacokinetic handling of the insecticide

(Knaak et al. 1965). However, a later study showed that intentionally dosed humans excreted only 25% –30% of carbaryl in urine after 24 hr, indicating that the fate of very significant fractions of the dose was unknown (Knaak et al. 1968).

Several studies estimating the degree to which carbaryl is absorbed through the skin has been previously reviewed by DPR (2006). Feldman and Maibach (1974) reported that 73.9% was absorbed by humans within 120 hr following an initial 24-hr dermal application of $^{14}$C-carbaryl at 4 μg/cm$^2$ (vehicle: acetone). Because other routes of disposition and excretion were not monitored, a 13.5-fold correction factor was imposed on the urinary value, based on the observation that only 7.4% of an i.v. dose appeared in the urine in 24 hr. However, the position of the $^{14}$C label was not reported. If, for example, the carbonyl or *N*-methyl carbon (as opposed to the naphthalene ring) was labeled, much of it would have been excreted as $CO_2$, resulting in an underestimate of urinary excretion and consequent overestimate of absorption. Shah and Guthrie (1983) determined carbaryl absorption at the same dermal dose of 4 μg/cm$^2$ (vehicle: acetone) applied to rat skin for up to 120 hr. Radiolabel in all tissues and excreta was measured, obviating the need for correction factors. Thus, they reported 72.1%, 75.1%, and 95.7% absorption at 12, 24, and 120 hr, respectively. Cheng (1995) also exposed rats to the insecticide by the dermal route for up to 24 hr, but at 35.6, 403, and 3,450 μg/cm$^2$. However, instead of acetone, aqueous carboxymethylcellulose was used as the vehicle. The author reported that absorption was inversely related to dose, with 24-hr values of 34%, 25%, and 4% with ascending dose. Ultimately, it appears that dose has a greater influence than vehicle choice in determining dermal carbaryl absorption (DPR 2006).

# 7 Summary

Carbaryl is an agricultural and garden insecticide that controls a broad spectrum of insects. Although moderately water soluble, it neither vaporizes nor volatilizes readily. However, upon spray application the insecticide is susceptible to drift. It is unstable under alkaline conditions, thus easily hydrolyzed. Carbaryl has been detected in water at ppb concentrations but degradation is relatively rapid, with 1-naphthol identified as the major degradation product. Indirect and direct photolysis of carbaryl produces different naphthoquinones as well as some hydroxyl substituted naphthoquinones.

Sorption of the insecticide to soil is kinetically rapid. However, although both the mineral and organic fractions contribute, because of its moderate water solubility it is only minimally sorbed. Also, sorption to soil minerals strongly depends on the presence of specific exchangeable cations and increases with organic matter aromaticity and age. Soil microbes (bacteria and fungi) are capable of degrading carbaryl; the process is more rapid in anoxic than aerobic systems and with increased temperature and moisture.

Carbaryl presents a significant problem to pregnant dogs and their offspring, but some have questioned the applicability of these data to humans. In addition, for toxicokinetic and/or physiological reasons, it has been argued that dogs are more sensitive than humans to carbaryl-induced reproductive or developmental toxicity. However, these arguments are based on either older pharmacokinetic studies or on speculation about possible reproductive differences between dogs on the one hand and rats and humans on the other. In view of the wider evidence from both human epidemiological and laboratory animal studies, the question of the possible developmental and reproductive toxicity of carbaryl should be considered open and requiring further study.

**Acknowledgments** Support was provided by the Environmental Monitoring Branch of the California Department of Pesticide Regulation (CDPR), California Environmental Protection Agency (CalEPA), and the California Rice Research Board. The statements and conclusions in this report are those of the authors and not necessarily those of the supporting agencies. The mention of commercial products, their source, or their use in connection with materials reported herein is not to be construed as actual or implied endorsement of such products.

# References

Ahmed FE, Lewis NJ, Hart RW (1977a) Pesticide induced ouabain resistant mutants in Chinese hamster V79 cells. Chem Biol Interact 19:269–374.
Ahmed FE, Hart RW, Lewis NJ (1977b) Pesticide induced DNA damage and its repair in cultured human cells. Mutat Res 42:161–174.
Ahmad R, Kookana RS, Alston AM, Bromilow RH (2001a) Differences in sorption behavior of carbaryl and phosalone in soils from Australia, Pakistan, and the United Kingdom. Aust J Soil Res 39:893–908.
Ahmad R, Kookana RS, Alston AM, Skjemstad JO (2001b) The nature of soil organic matter affects sorption of pesticides. 1. Relationships with carbon chemistry as determined by $^{13}$C CPMAS NMR spectroscopy. Environ Sci Technol 35:878–884.
Arroyo LJ, Li H, Teppen BJ, Johnston CT, Boyd SA (2004) Hydrolysis of carbaryl by carbonate impurities in reference clay SWy-2. J Agric Food Chem 52:8066–8073.
Austin EW (2002) 4 week repeated-dose dermal toxicity study with carbaryl technical in rats. Covance Laboratories Inc. Lab project no. 6224-268, DPR vol no. 169–413, rec no. 186206.
Baron RL (1991) Carbamate insecticides. In: Hayes WJ Jr, Laws ER Jr (eds) Handbook of Pesticide Toxicology, vol 3. Academic Press, San Diego, pp 1125–1189.
Beyrouty P (1992) An acute study of the potential effects of orally administered carbaryl on behavior in rats. Bio-Research Laboratories Ltd. Lab project no. 97109.
Bondarenko S, Gan J (2004) Degradation and sorption of selected organophosphate and carbamate insecticides in urban steam sediments. Environ Toxicol Chem 23: 1809–1814.
Brahmia O, Richard C (2003) Phototransformation of carbaryl in aqueous solution: laser-flash photolysis and steady-state studies. J Photochem Photobiol A 156:9–14.
Branch RA, Jacqz E (1986) Subacute neurotoxicity following long-term exposure to carbaryl. Am J Med 80:741–745.
Brodeur J, DuBois KP (1963) Comparison of acute toxicity of anticholinesterase insecticides to weanling and adult male rats. Proc Soc Exp Biol Med 114:509–511.
Brooks W, Broxup B (1995) A time of peak effects study of a single orally administered dose of carbaryl, technical grade, in rats. Bio-Research Laboratories Ltd. Lab project no. 97388, DPR vol no. 169–338, rec no. 142593.
Brooks W, Robinson K, Broxup B (1995) An acute study of the potential effects of a single orally administered dose of carbaryl, technical grade, on behavior and neuromorphology in rats. Bio-Research Laboratories Ltd. Lab project no. 97389, DPR vol no. 169–341, rec. no. 142602.

Carpenter CP, Weil CS, Palm PE, Woodside MW, Nair JH, Smyth HF (1961) Mammalian toxicity of 1-naphthyl-$N$-methylcarbamate (sevin insecticide). J Agric Food Chem 9:30–39.
CDPR (2005) Summary Report. Complete Report with Summary Data Indexed by Chemical. California Environmental Protection Agency, Department of Pesticide Regulations, Sacramento, CA, pp 340–341. http://www.cdpr.ca.gov/docs/pur/pur04rep/chmrpt04.pdf (verified August 10, 2007).
Cheng T (1995) Dermal absorption of $^{14}$C-carbaryl (XLR Plus) in male rats (preliminary and definitive phases). Rhone-Poulenc Agro. Lab project no. HWI-6224-206, DPR vol no. 169–386, rec no. 167057.
Cifone MA (1989) Mutagenicity test on carbaryl technical in the in vitro rat primary hepatocyte unscheduled DNA synthesis assay. Hazleton Laboratories America. Study no. 10862-0-447, DPR vol no. 169–196, rec no. 85659.
Collins TFX, Hansen WH, Keeler HV (1971) The effect of carbaryl (sevin) on reproduction of the rat and the gerbil. Toxicol Appl Pharm 19:202–216.
Cranmer MF (1986) Carbaryl: a toxicological review and risk analysis. Neurotoxicology 7:247–332.
Cox C (1993) The problems with Sevin (carbaryl). J Pestic Reform 13:31–36.
Currier W, MacCollom G, Baumann G (1982) Drift residues of air-applied carbaryl in an orchard environment. J Econ Entomol 75:1062–1068.
De Mel GM, Pathiratne A (2005) Toxicity assessment of insecticides commonly used in rice pest management to the fry of common carp, *Cyprinus carpio*, a food fish culturable in rice fields. J Appl Ichthyol 21:146–150.
De Oliveira MF, Johnston CT, Premachandra GS, Teppen BJ, Li H, Laird DA, Zhu D, Boyd SA (2005) Spectroscopic study of carbaryl sorption on smectite from aqueous suspension. Environ Sci Technol 39:9123–9129.
Desi I, Gonczi L, Simon G, Farkas I, Kneffel Z (1974) Neurotoxicologic studies of two carbamate pesticides in subacute animal experiments. Toxicol Appl Pharm 27:465–476.
Deutsch-Wenzel RP, Brune H, Grimmer G, Misfeld J (1985) Local application to mouse skin as a carcinogen specific test system for non-volatile nitroso-compounds. Cancer Lett 29:85–92.
Dickoff DJ, Gerber O, Turovsky Z (1987) Delayed neurotoxicity after ingestion of carbamate pesticide. Neurology 37:1229–1231.
Doddamani HP, Ninnekar HZ (2001) Biodegradation of carbaryl by a micrococcus species. Curr Microbiol 43:69–73.
Dorough HW, Casida JH (1964) Nature of certain carbamate metabolites of the insecticide Sevin. J Agric Food Chem 12:294–304.
DPR (2006) Dermal absorption of carbaryl (memo from S. Beauvais to J.P. Frank). Worker Health and Safety Branch, California Environmental Protection Agency, Department of Pesticide Regulation, Sacramento, CA.
Dudek R (1985) Four hour acute dust inhalation toxicity study in rats of sevin 97.5 MC. American Biogenics. Study no. 420-1979, DPR vol no. 169-114, rec. no. 51935.
Eichelberger JW, Lichtenberg JJ (1971) Persistence of pesticides in river water. Environ Sci Technol 5:541–544.
Eisenbrand G, Ungerer O, Preussmann R (1975) The reaction of nitrite with pesticides. II. Formation, chemical properties and carcingenic activity of the N-nitroso derivative of $N$-methyl-1-naphthyl carbamte (carbaryl). Food Cosmet Toxicol 13:365–367.
Feldman RJ, Maibach HI (1974) Percutaneous penetration of some pesticides and herbicides in man. Toxicol Appl Pharmacol 28:126–132.
Ferrari A, Anguiano OL, Venturion A, Pechen de D'Angelo AM (2004) Different susceptibility of two aquatic vertebrates (*Oncorhynchus mykiss* and *Bufo arenarum*) to azinphos methyl and carbaryl. Comp Biochem Physiol 139:239–243.
Foreman WT, Majewski MS, Goolsby DA, Wiebe FW, Coupe RH (2000) Pesticides in the atmosphere of the Mississippi River Valley, part II: air. Sci Total Environ 248:213–216.
Fytianos K, Pitarakis K, Bobola E (2006) Monitoring of $N$-methylcarbamate pesticides in the Pinios River (central Greece) by HPLC. Int J Environ Anal Chem 86:131–145.

Ghauch A, Gallet C, Charef A, Rima J, Martin-Bouyer M (2001) Reductive degradation of carbaryl in water by zero-valent iron. Chemosphere 42:419–424.

Gilliom RJ, Barbash JE, Crawford CG, Hamilton PA, Martin JD, Nakagaki N, Nowell LH, Scott JC, Stackelberg PE, Thelin GP, Wolock DM (2007) The quality of our nation's waters; pesticides in the nation's streams and groundwater, 1992–2001. U.S. Geological Survey Circular 1291.

Grover IS, Ladhar SS, Randhawa SK (1989) Carbaryl: a selective genotoxicant. Environ Pollut 58:313–323.

Hamada NH (1987) One-year oral toxicity study in beagle dogs with carbaryl technical. Hazleton Laboratories America. Lab project no. 400-715, DPR vol no. 169-169, rec. no. 056429.

Hamada NH (1993a) Oncogenicity study with carbaryl technical in CD-1 mice. Hazleton Laboratories America. Lab project no. 656-138, DPR vol no. 169–267, rec. no. 123769.

Hamada NH (1993b) Combined chronic toxicity and oncogenicity study with carbaryl technical in Sprague-Dawley rats. Hazleton Laboratories America. Lab project no. 656-139, DPR vol no. 169–271, rec. no. 126241.

Hastings FL, Holsten EH, Shea PJ, Werner RA (2001) Carbaryl: a review of its use against bark beetles in coniferous forests of north America. Environ Entomol 30:803–810.

Holbert MS (1989) Acute inhalation toxicity study in rats. Stillmeadow. Study no. 6280-89, DPR vol no. 169–238, rec no. 97885.

Ishidate M, Odashima S (1977) Chromosome tests with 134 compounds on Chinese hamster cells in vitro: a screening for chemical carcinogens. Mutat Res 48:337–354.

Jana T, Das B (1997) Sorption of carbaryl (1-napthyl $n$-methyl carbamate) by soil. Bull Environ Contam Toxicol 59:65–71.

Kao AS (1994) Formation and removal reactions of hazardous air pollutants. J Air Waste Manag 44:683–696.

Karinen JF, Lambertson JG, Stewart NE, Terrier LC (1967) Persistence of carbaryl in marine estuarine systems. J Agric Food Chem 15:148.

Knaak JB, Sullivan LJ (1967) Metabolism of carbaryl in the dog. J Agric Food Chem 15:1125–1126.

Knaak JB, Tallant MJ, Bartley WJ, Sullivan LJ (1965) The metabolism of carbaryl in the rat, guinea pig, and man. J Agric Food Chem 13:537–543.

Knaak JB, Tallant MJ, Kozbelt SJ, Sullivan LJ (1968) The metabolism of carbaryl in man, monkey, pig, and sheep. J Agric Food Chem 16:465–470.

Krolski ME, Nguyen T, Lopez R, Ying L-L, Roensch W (2003) Metabolism and pharmacokinetics of [$^{14}$C]-carbaryl in rats following mixed oral and dermal exposure. Bayer CropScience. Study no. 04MECAY004, DPR vol no. 169-0476, rec no. 225213.

LaFleur K (1976) Movement of carbaryl through congaree soil into ground water. J Environ Qual 5:91–92.

Larson DM (1987a) Acute dermal toxicity evaluation of Carbaryl 90DF in rabbits. Toxicology Pathology Services. Study 320C-301-210-87; DPR vol 169–238, rec. 97888.

Larson DM (1987b) Acute oral toxicity evaluation of carbaryl 90DF in rats. Toxicology Pathology Services. Study 320A-101-010-87, DPR vol no. 169–238, rec no. 97883.

Larson DM (1987c) Evaluation of the sensitization potential of carbaryl 90DF in guinea pigs. Toxicology Pathology Services. Study no. 320B-201-215-87, DPR vol no. 169–238, rec. no. 97884.

Lawlor TE (1989) Mutagenicity test on carbaryl (technical) in the Ames *Salmonella* microsome reverse mutation assay. Hazleton Laboratories America. Study no. 10862-0-401, DPR vol no. 169–196, rec no. 85660.

Li AY, Davey RB, George JE (2005) Carbaryl resistance in Mexican strains of the southern cattle tick (Acari: Ixodidae). J Econ Entomol 98:552–556.

Lijinsky W, Schmahl D (1978) Carcinogenicity of $n$-nitroso derivatives of $n$-methylcarbamate insecticides in rats. Ecotoxicol Environ Saf 2:413–419.

Lijinsky W, Taylor HW (1976) Carcinogenesis in Sprague-Dawley rats of n-nitrosos-n-alkylcarbamate esters. Cancer Lett 1:275–279.

Marshall R (1996) Carbaryl: induction of micronuclei in the bone marrow of treated mice. Corning Hazleton. Study no. 198/89-1052, DPR vol no. 169–458, rec no. 209661.

Martin JD, Crawford CG, Larson SJ (2003) Pesticides in streams: preliminary results from cycle I of the National Water Quality Assessment Program (NAWQA), 1992–2001.
Meeker JD, Ryan L, Barr DB, Herrick RF, Bennett DH, Bravo R, Hauser R (2004a) The relationship of urinary metabolites of carbaryl/naphthalene and chlorpyrifos with human semen quality. Environ Health Perspect 112:1665–1670.
Meeker JD, Singh NP, Ryan L, Duty SM, Barr DB, Herrick RF, Bennett DH, Hauser R (2004b) Urinary levels of insecticide metabolites and DNA damage in human sperm. Hum Reprod 19:2573–2580
Meister RT (2001) Farm Chemicals Handbook. Meister Publishing Company, Willoughby, OH.
Mellon Inst (1957) The toxicity of insecticide sevin. Report 20–89, DPR vol no. 169-055, rec. no. 23486.
Menon P, Gopal M (2003) Dissipation of $^{14}$C carbaryl and quinalphos in soil under a groundnut crop (*Arachis hypogaea* L.) in semi-arid India. Chemosphere 53:1023–1031.
Miller PL, Chin Y-P (2002) Photoinduced degradation of carbaryl in wetland surface water. J Agric Food Chem 50:6758–6765.
Murray FJ, Staples RE, Schwetz BA (1979) Teratogenic potential of carbaryl given to rabbits and mice by gavage or by dietary inclusion. Toxicol Appl Pharmacol 51:81–89.
Murli H (1989) Mutagenicity test on carbaryl technical in an *in vitro* cytogenetic assay measuring chromosomal aberration frequencies in Chinese hamster ovary (CHO) cells. Hazleton Laboratories America. Study no. 10862-0-437. DPR vol no. 169–196, rec. no. 85657.
MG, Kavlock RJ (1995) A multidisciplinary approach to toxicological screening: II. Developmental toxicity. J Toxicol Environ Health 45:145–171.
Onfelt A, Klasterska I (1984) Sister-chromatid exchanges and thioguanine resistance in V79 Chinese hamster cells after treatment with the aneuploidy-inducing agent carbaryl ± S9 mix. Mutat Res 125:269–274.
Osman M, Belal M (1980) Persistence of carbaryl in canal water. J Environ Sci Health B 15:307–311.
Pant N, Srivastava SC, Prasad AK, Shankar R, Srivastava SP (1995) Effects of carbaryl on the rat's male reproductive system. Vet Hum Toxicol 37:421–425.
Pant N, Shankar R, Srivastava SP (1996) Spermatotoxic effects of carbaryl in rats. Hum Exp Toxicol 15:736–738.
Phillips PJ, Bode RW (2004) Pesticides in surface water runoff in south-eastern New York State, USA: seasonal and stormflow effects on concentrations. Pestic Manag Sci 60:531–543.
Rajagopal BS, Chendrayan K, Reddy BR, Sethunathen N (1983) Persistence of carbaryl in flooded soils and its degradation by soil enrichment cultures. Plant Soil 73:35–45.
Rajagopal BS, Brahmaprakash GP, Reddy BR, Singh UD, Sethunathan N (1984) Effect and persistence of selected carbamate pesticides in soils. In: Gunther FA, Gunter JD (eds) Reviews of Environmental Contamination and Toxicology. Residue Rev 93:87–203.
Regan JD, Setlow RB, Francis AA, Lijinsky W (1976) Nitrosocarbaryl: its effect on human DNA. Mutat Res 38:293–302.
Repetto-Larsay M (1998) Carbaryl: developmental toxicology study in the rat by gavage. Rhone-Poulenc Agro. Study no. SA 98070, DPR vol no. 169–383, rec. no. 166125.
Reyes JGG, Leyva NR, Millan OA, Lazcano GA (2002) Effects of pesticides on DNA and protein of shrimp larvae *Litopenaeus stylirostris* of the California Gulf. Ecotoxicol Environ Saf 53:191–195.
Ribera D, Narbonne JF, Arnaud C, Denis MS (2001) Biochemical responses of the earthworm *Eisenia fetida andrei* exposed to contaminated artificial soil: effects of carbaryl. Soil Biol Biochem 33:1123–1130.
Rickard RW, Dorough HW (1984) In vitro formation of nitrosocarbamates in the stomach of rats and guinea pigs. J Toxicol Environ Health 142:279–290.
Robinson K, Broxup B (1997) A developmental neurotoxicity study of orally administered carbaryl, technical grade, in the rat. ClinTrials BioResearch Ltd. Lab project no. 97391, DPR vol no. 169–384, rec. no. 166126.
Rodriguez LD, Dorough HW (1977) Degradation of carbaryl by soil microorganisms. Arch Environ Contam Toxicol 6:47–56.
Rybakova MN (1966) Toxic effect of sevin on animals. Gigena Sanitariya 31:402–407.

Sagelsdorff P (1994) Investigation of the potential for protein- and DNA-binding of carbaryl. CIBA-GEIGY Ltd. Lab project no. CB93/52, DPR vol no. 169–456, rec. no. 209659.

Sakamoto M, Chang KH, Hanazato T (2005) Differential sensitive of a predacious cladoceran (*Leptodora*) and its prey (the cladoceren *Bosmina*) to the insecticide carbaryl: results of acute toxicity tests. Bull Environ Contam Toxicol 75:28–33.

Sanusi A, Millet M, Mirabel P, Wortham H (2000) Comparison of atmospheric pesticide concentrations measured at three sampling sites: local, regional and long-range transport. Sci Total Environ 263:263–277.

Savitz DA, Arbuckle T, Kaczor D, Curtis KM (1997) Male pesticide exposure and pregnancy outcome. Am J Epidemiol 146:1025–1036.

Shah PV, Guthrie FE (1983) Percutaneous penetration of three insecticides in rats: a comparison of two methods for in vivo determination. J Invest Dermatol 80:291–293.

Sharom MS, Miles JR, Harris CR, McEwen FL (1980) Behavior of 12 insecticides in soil and aqueous suspensions of soil and sediment. Water Res 14:1095–1100.

Shehata T, Eichardson E, Cotton E (1984) Assessment of human population exposure to carbaryl from the 1982 main spruce budworm spray project. J Environ Health 46:293–297.

Sheng G, Johnston CT, Teppen BJ, Boyd SA (2001) Potential contributions of smectite clays and organic matter to pesticide retention in soils. J Agric Food Chem 49:2899–2907.

Shtenberg AI, Rybakova MN (1968) Effect of carbaryl on the neuroendocrine system of rats. Food Cosmet Toxicol 6:461–467.

Smalley HE, Curtis JM, Earl FL (1968) Teratogenic action of carbaryl in beagle dogs. Toxicol Appl Pharm 13:392–403.

Soderpalm-Berndes C, Onfelt A (1988) The action of carbaryl and its metabolite α-naphthol on mitosis in V79 Chinese hamster fibroblasts. Indications of the involvement of some cholinester in cell division. Mutat Res 201:349–363.

Sogorb MA, Carrera V, Benabent M, Vilanova E (2002) Rabbit serum albumin hydrolyzes the carbamate carbaryl. Chem Res Toxicol 15:520–526.

Sogorb MA, Carrera V, Vilanova E (2004) Hydrolysis of carbaryl by human serum albumin. Arch Toxicol 78:629–634.

Stenersen J (1992) Uptake and metabolism of xenobiotics by earthworms. In: Greigh-Smith PW, Becker H, Edwards PJ, Heimbach F (eds) Ecotoxicology of Earthworms. Intercepts Ltd, Andover, UK, pp 129–138.

Struble CS (1994) Metabolism of $^{14}$C-carbaryl in rats (preliminary and definitive phases). Hazleton Wisconsin Inc. Lab project no. HWI 6224-184, DPR vol no. 169–453, rec. no. 209656.

Sun F, Zhu T, Shang I, Han L (2005) Gas-phase reaction of dichlorvos, carbaryl, chlordimeform, and 2,4-D butyl ester with OH radicals. Int J Chem Kinet 37:755–762.

Tang J, Cao Y, Rose RL, Hodgson E (2002) In vitro metabolism of carbaryl by human cytochrome p450 and its inhibition by chlorpyrifos. Chem Biol 141:229–241.

Todd NE, Van Leeuwan MV (2002) Effects of sevin (carbaryl insecticide) on early life stages of zebrafish (*Danio rerio*). Ecotoxicol Environ Saf 53:267–272.

Tomlin CDS (2000) The Pesticide Manual, 12th Ed. British Crop Protection Council, Surrey, UK, pp 133–134.

Tomlin CDS (2003) The Pesticide Manual, 13th Ed. British Crop Protection Council, Surrey, UK, pp 135–136.

Totis M (1997) Investigation of the metabolism of $^{14}$C-carbaryl in the 15 month old male rat following chronic dietary administration. Rhone-Poulenc Acrochimie. Lab project no. SA 95288, DPR vol no. 169–454, rec. no. 209657.

Tripathi PK, Singh A (2001) Toxic effects of dimethoate and carbaryl pesticides on carbohydrate metabolism of freshwater snail *Lymnaea acuminate*. Bull Environ Contam Toxicol 70:717–722.

Tripathi PK, Singh A (2004) Carbaryl induced alteration in the reproduction and metabolism of freshwater snail *Lymnaea acuminata*. Pestic Biochem Phys 79:1–9.

Tyl RW, Marr MC, Myers CB (1999) Developmental toxicity evaluation (with cholinesterase assessment) of carbaryl administered by gavage to New Zealand white rabbits. Research Triangle Institute. Report no. 65C-7297-200/100, DPR vol no. 169–389, rec. no. 170646.

Tyl RW, Myers CB, Marr MC (2001) Two-generation reproductive toxicity evaluation of carbaryl (RPA007744) administered in the feed to CD (Sprague-Dawley) rats. Research Triangle Institute. Report no. 65C-07407-400, DPR vol no. 169–410, rec no. 182115.

Union Carbide (1983a) Sevin 99% Technical: acute toxicity and irritancy study. Project no. 46-71, DPR vol no. 169-085, rec. no. 695 (oral toxicity).

Union Carbide (1983b) Sevin 99% Technical: acute toxicity and irritancy study. Project no. 46-71, DPR vol no. 169-085, rec. no. 693 (dermal toxicity).

Union Carbide (1983c) Sevin 99% Technical: acute toxicity and irritancy study. Project no. 46-71, DPR vol no. 169-085, rec. no. 694 (primary eye irritation).

Union Carbide (1983d) Sevin 99% Technical: acute toxicity and irritancy study. Project no. 46–71, DPR vol no. 169-085, rec. no. 696 (primary dermal irritation).

Union Carbide (1985) Compilation of available APC carbaryl acute toxicity data from 1954 to the present. DPR vol no. 169-117, rec no. 34482.

US EPA (2002) Carbaryl: Updated Toxicology Chapter for RED (Dobozy VA). DP Barcode D282980.

US EPA (2003) Carbaryl: Revised Dietary Exposure Analysis Including Acute Probabilistic Water Analysis for the HED Revised Human Health Risk Assessment (Fort FA). Chemical no. 056801/ List A Re-registration Case no. 0080. DP Barcode D288479.

US EPA (2004) Interim Re-registration Eligibility Document for Carbaryl. List A, Case 0080. Revision date: October 22, 2004.

USGS (2007) NAWQA data warehouse. United States Geological Survey. Available at: http://infotrek.er.usgs.gov/traverse/f?p=NAWQA:HOME (verified August 17, 2007).

Valles B (1999) Investigation of the metabolism of $^{14}C$-carbaryl following 14 days administration to the male $CD_1$ mouse. Rhone-Poulenc Agro. Lab project no. SA 97481, DPR vol no. 169–402, rec. no. 177759.

Venkateswarlu K, Chendrayan K, Sethunathan N (1980) Persistence and biodegradation of carbaryl in soils. J Environ Sci Health B 15:421–429.

Walters J, Goh KS, Li L, Feng H, Hernandez J, White J (2003) Environmental monitoring of carbaryl applied in urban areas to control the glassy-winged sharpshooter in California. Environ Monit Assess 82:265–280.

Wang Q, Lemley AT (2002) Oxidation of carbaryl in aqueous solution by membrane anodic fenton treatment. J Agric Food Chem 50:2331–2337.

Ware GW (2000) The Pesticide Book. Thomson Publications, Fresno, CA, pp 57, 83, 302.

Weil CJ (1972) Comparative study of dietary inclusion versus stomach intubation on three-generations of reproduction, on teratology and on mutagenesis. Mellon Institute. Report no. 35–65, DPR vol no. 169-099, rec. no. 27202.

Wilson PC, Foos JF (2006) Survey of carbamate and organophosphorous pesticide export from a south Florida (USA) agricultural watershed: implications of sampling frequency on ecological risk estimation. Environ Toxicol Chem 25:2847–2852.

Wolfe NL, Zepp RC, Paris DF (1978) Carbaryl, propham, chloropropham: a comparison of rates of hydrolysis and photolysis with the rate of biolysis. Water Res 12:565.

Wyrobek AJ, Watchmaker G, Gordon L, Wong K, Moore D, Whorton D (1981) Sperm shape abnormalities in carbaryl-exposed employees. Environ Health Perspect 40:255–265.

Xia Y, Cheng S, Bian Q, Xu L, Collins MD, Chang HC, Song L, Liu J, Wang S, Wang X (2005) Genotoxic effects on spermatozoa of 1carbaryl-exposed workers. Toxicol Sci 85:615–623.

Young RR (1989) Mutagenicity test on carbaryl (technical) in the CHO/HGPRT forward mutation assay. Hazleton Laboratories America. Study no. 10862-0-435, DPR vol no. 169-106, rec. no. 85658.

# Managing Hazardous Pollutants in Chile: Arsenic

Ana María Sancha and Raul O'Ryan

## Contents

1 Introduction ............................................................................................................ 123
2 Arsenic Contamination in Chile ............................................................................ 125
  2.1 Arsenic Levels in Air, Water, Soil, and Food ................................................ 125
  2.2 Population Exposure ...................................................................................... 129
  2.3 Health Effects ................................................................................................. 131
3 Removing Arsenic from Water .............................................................................. 132
  3.1 Arsenic Removal Technologies used in Chile ............................................... 132
  3.2 Costs ............................................................................................................... 135
4 Reducing Airborne Arsenic Emissions ................................................................. 137
  4.1 Control Options .............................................................................................. 138
  4.2 Costs to Reduce Arsenic Air Pollution .......................................................... 139
5 Setting Environmental Standards for Arsenic: Present and Future ...................... 140
  5.1 Air ................................................................................................................... 140
  5.2 Water .............................................................................................................. 141
6 Concluding Remarks ............................................................................................. 142
7 Summary ............................................................................................................... 143
References ................................................................................................................. 144

## 1 Introduction

Chile is one of the few countries confronted with environmental challenges posed by extensive arsenic pollution, which exists in the northern part of the country. Naturally occurring arsenic in Chile derives from volcanic activity in the Andes Mountains and affects water, air, and soils. Additionally, copper mining and smelting activities, major economic activities in Chile, are important anthropogenic sources

---

A.M. Sancha
Civil Engineering Department, University of Chile, Blanco Encalada 2002, Santiago, Chile
e-mail: amsancha@ing.uchile.cl

R. O'Ryan
Industrial Engineering Department and Center for Applied Economics, University of Chile, Santiago, Chile

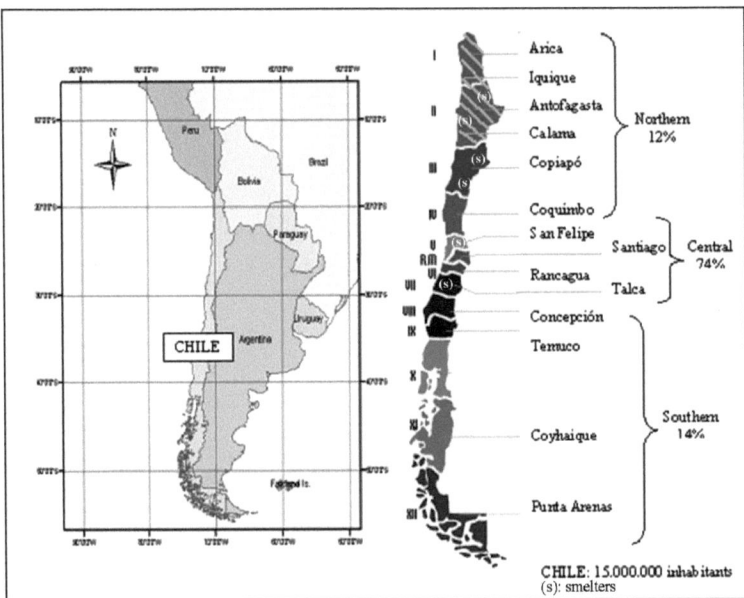

**Fig. 1** Geographic distribution of population (% by zone), main cities, and smelters in Chile

of arsenic. The high levels of arsenic contamination in the north of Chile, and the economic consequences of mitigating the contamination in water and air, have not allowed copying standards applied in other countries.

In Chile, approximately 1.8 million people, representing about 12% of the total population, live in arsenic-polluted areas (Fig. 1). Until recently, water consumed by the urban population contained levels of arsenic that were much higher than the values recommended by the World Health Organization (WHO). The air near many large cities is also contaminated with arsenic in variable amounts, which derives from both natural sources and intensive mining activity near those cities. In rural areas, indigenous populations are exposed to arsenic by consuming polluted water and various foods grown locally in arsenic-contaminated soils.

The health effects of arsenic were first discovered in the 1950s, when vascular, respiratory, and skin lesions from drinking arsenic-contaminated drinking water were observed in children and adults. Methodologies were evaluated and technology developed to mitigate arsenic in polluted water sources, which was accomplished at a relatively low cost. Abatement began in the 1970s, with construction and operation of treatment plants designed to remove arsenic from contaminated water supplies. An exacerbating factor is that water is extremely scarce in northern Chile, where the Atacama Desert exists; this desert is among the driest in the world.

During the 1980s and early 1990s, epidemiological studies showed that the rate of lung and bladder cancer in the northern zone was considerably higher than the mean cancer rate of the general Chilean population. Exposure to arsenic was regarded to be the source of this increased incidence of cancer.

In the 1990s, the presence of high arsenic levels in the vicinity of copper smelters gave rise to action by Chile's Health Ministry to regulate arsenic smelter emissions.

The measures undertaken to reduce atmospheric arsenic levels by regulating emissions from copper smelting plants in northern Chile were prompted, to some extent, by international pressures and concerns.

The process of establishing emission standards to abate arsenic was unique for a developing country. The process included an extensive evaluation of both risks and costs for various abatement options. The evaluation was challenging because of a general lack of systematic information on health effects, emission and contamination levels, control technologies, and related costs. To address this dearth of information, a research project entitled "Protection of the Competitiveness of Chile's Mining Products: Antecedents and Criteria for Environmental Regulation of Arsenic" (FONDEF 2–24 1997) was undertaken from 1994 to 1996. In this project, the independent Universidad de Chile, working with the Ministries and mining and sanitary companies involved, developed the required information. As a result, considerable experience was gained on how to manage this type of hazardous pollution, particularly in the context of evaluating trade-offs among production, control costs, and health improvement.

More recently, the local community has pushed for more significant reductions of arsenic concentrations in air and water, mainly because of their increasing awareness and knowledge concerning risks from exposure to arsenic.

In this chapter, we present a historical assessment of arsenic contamination in Chile. Levels of arsenic in air, water, soils, and vegetables, together with the implications to health and patterns of exposure, are reviewed. Moreover, we relate the important lessons learned on how a developing country approaches regulation of a hazardous pollutant in the face of scarce financial, technical, and human resources.

## 2 Arsenic Contamination in Chile

### 2.1 Arsenic Levels in Air, Water, Soil, and Food

**Background Levels**

Knowledge of background levels of a particular contaminant in its ambient state is a fundamental requirement before the pollutant can be regulated. In Chile, the regional geology and extent of mining activities determine what relative environmental arsenic levels are likely to be encountered in Chile's northern, central, and southern zones (Enriquez 1978). In this section we review what is known of arsenic levels in Chile's air, water, soil, and food.

**Air**

The literature provides scant information on arsenic levels in air, regardless of country, although data from work sites are available (EHC 1981). A detailed study of airborne arsenic concentrations in principal Chilean cities was made during the period 1994–1995 (Ulriksen and Cabello 2004). Levels of breathable particulate airborne arsenic (PM-10) were determined by taking 24-hr samples for a 7-d period

**Table 1** Concentrations of arsenic (As) in atmospheric particulate matter for selected Chilean cities

| Cities | Period (yr) | No. of samples | Concentration ($\mu g/m^3$) | |
|---|---|---|---|---|
| | | | Average | Maximum |
| Northern zone | | | | |
| Arica | 01/96–02-96 | 10 | 0.004 | 0.007 |
| Iquique | 06/94–06/95 | 85 | 0.025 | 0.037 |
| Antofagasta | 06/94–06/95 | 93 | 0.057 | 0.088 |
| Central zone | | | | |
| Viña del Mar | 06/94–03/95 | 84 | 0.026 | 0.051 |
| Santiago | 07/94–06/95 | 83 | 0.02 | 0.037 |
| Rancagua | 06/94–06/95 | 85 | 0.038 | 0.081 |
| Talca | 06/94–06/95 | 84 | 0.003 | 0.007 |
| Southern zone | | | | |
| Concepción | 06/94–06/95 | 84 | 0.007 | 0.016 |

once each month for an entire year. These measurements were designed to determine average annual values, seasonal variations, and day-to-day variations while optimizing use of resources and minimizing monitoring and analysis costs. In the southern zone, samples were taken only at the city of Concepción, because it is known that, in this region, neither natural geological nor anthropogenic sources of arsenic contamination exist. Table 1 presents the results of this airborne arsenic contamination study.

Sampling results indicate that the lowest levels of arsenic were found in Arica, Talca, and Concepción, cities relatively distant from mining activities. The highest arsenic concentrations were registered at Antofagasta, a city in the northern zone, where geological sources of arsenic have been exacerbated by anthropogenic contamination from smelting emissions.

In the central zone, the city of Rancagua also registered high airborne arsenic levels from nearby copper smelting activity. Although distant from such activity, Santiago recorded midrange arsenic levels, influenced both by combustion of coal and petroleum and, probably, by other unidentified industrial sources.

The reported analytical results for arsenic have the usual sources of uncertainty. Other than the standard sampling uncertainty, there are variations over time, both in the mineral composition of the ore smelted and in patterns of production utilized, as well as in seasonal variability and meteorological factors. However, for cities unaffected by mining activity, we presume reported airborne arsenic levels reasonably represent actual historic levels of arsenic pollution.

## Water

The natural quality of water in Chile is closely related to the geochemical characteristics of the soils and rocks with which it comes into contact. Table 2 shows arsenic levels that exist in a variety of rivers in Chile. These results show that

Table 2 Arsenic concentrations in water of selected Chilean rivers, by geographic zone

| Rivers | Arsenic range (µg/L) |
|---|---|
| Northern zone | |
| San José | 50–100 |
| Lluta | 600–700 |
| Toconce | 600–900 |
| Lequena | 150–350 |
| Colana | 70–90 |
| Siloli | 20–50 |
| Inacaliri | 80–90 |
| Central zone | |
| Elqui | 50–60 |
| Blanco | 20–30 |
| Aconcagua | 10–20 |
| Maipú | 10–20 |
| Mapocho | 10–20 |
| Southern zone | |
| Bío-Bío | <5 |
| Valdivia | <5 |

hydrological resources in the north of Chile generally have elevated, if variable, arsenic levels. To satisfy water quality criteria, these rather high levels require water treatment before consumption by urban residents of this zone. However, among indigenous peoples of the Altiplano, most still consume untreated waters (Sancha et al. 1995, 1997; Queirolo et al. 2000a).

Studies of arsenic speciation in water indicate that arsenic generally exists as $As^+$ (V) (Sancha et al. 1992a,b). In addition to arsenic, many northern waterways also contain boron (B), an element whose deleterious effects on public health is only recently beginning to be explored. Both elements are of concern to regional authorities.

## Soils and Vegetables

Several studies (De Gregori et al. 2003, 2004; Queirolo et al. 2000b) demonstrate that a wide range of arsenic concentrations are found in Chilean soils (Fig. 2). Elevated arsenic levels are encountered in nearly all soils in the northern zone. Such contamination is of geological origin and is distributed and enhanced through crop irrigation, and it is also a by-product of mining activity.

Studies of arsenic residues in vegetables and fruit sold in farmers' markets and grocery stores reveal significant concentration differences among zones (Fig. 3). High arsenic levels were found in irrigated vegetables and in soils of indigenous communities of the Altiplano (Table 3). It was noted that the concentration of

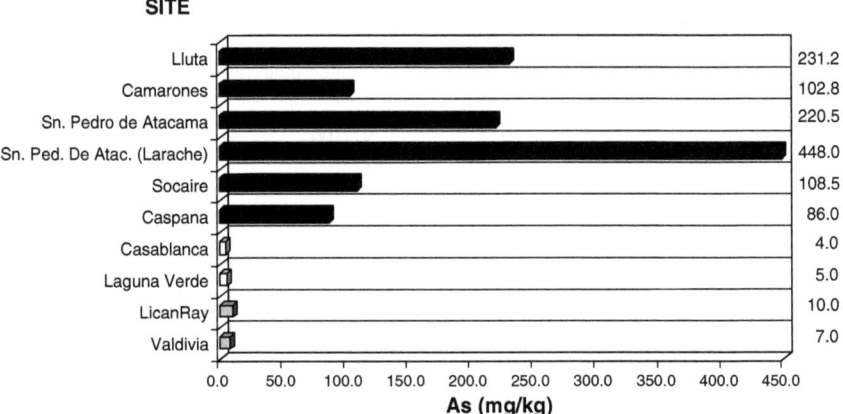

**Fig. 2** Arsenic levels in soils from various Chilean cities

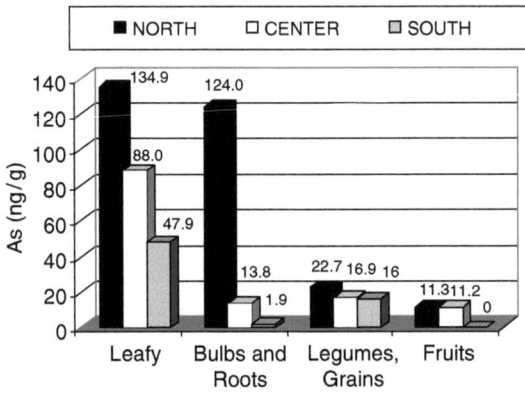

**Fig. 3** Arsenic in vegetables and fruits grown in Chile. Leafy vegetables include Swiss chard, lettuce, parsley, spinach, etc. Bulbs and roots include garlic, beets, carrots, onions, potatoes, etc. Legumes and grains include beans, corn, peas, etc. Fruits include apples, pears, grapes, and peaches

**Table 3** Arsenic concentrations in water, soil, and vegetables cultivated in the Altiplano

| Site[a] | Water (As µg/L) | Soil (As µg/g) | Vegetables (As µg/g) | | | | | |
|---|---|---|---|---|---|---|---|---|
| | | | Cabbage | Radish | Beet | Chard | Potato | Garlic | Onion |
| 1 | 2 | 86.0 | – | – | – | 0.218 | – | 0.018 | 0.036 |
| 2 | 172 | 220.5 | 0.054 | 0.207 | – | 0.282 | 0.040 | – | 0.106 |
| 3 | 220 | 108.5 | 0.033 | – | 0.156 | – | 0.044 | 0.050 | – |
| 4 | 619 | 448.0 | 0.715 | 0.938 | 0.520 | 0.718 | – | – | – |

[a] *Sites*: 1, Caspana; 2, San Pedro de Atacama (Condeduque); 3, Socaire; 4, San Pedro de Atacama (Larache).

arsenic in the same crop varieties depended on arsenic levels in irrigation water and soil in which these crops come into contact (Sancha et al. 1995). Although these results are interesting, they do not define a clear cause-and-effect relationship.

Agricultural production in such Altiplano communities is very limited because of high water and soil salinity, characteristic of the zone. For this reason, crops grown here are consumed locally and do not reach the larger cities of Chile's north.

Studies of arsenic levels in vegetables from As-endemic areas have increased in recent years (Queirolo et al. 2000a; Muñoz et al. 2002; Díaz et al. 2004). However, comparisons of results from one study to another are challenging, because, as some studies point out, arsenic levels found may depend on bioavailability of arsenic in soil (Flynn et al. 2002; Roychowdhury et al. 2002, 2003; Alam et al. 2003). Such bioavailability is known to be affected by soil pH, organic matter content, clay content, and clay type, among other variables. Another factor that may affect residue levels is the irrigation method utilized; drainage that cleanses soil could diminish bioavailability (Ayers and Westcot 1994).

## 2.2 Population Exposure

Chile's particular geological conditions and anthropogenic activities result in higher levels of environmental exposure to arsenic than levels considered normal in the literature. This is particularly true in Chile's northernmost region, but less so in the central zone (see Fig. 1). The cities of Antofagasta and Calama are especially affected by this contaminant.

Early estimates of arsenic exposure to the Chilean population were obtained in the 1990s. These data opened the way for evaluating the relative contribution to human arsenic exposure provided by various environmental media (air, water, and food) (Sancha et al. 1997; Sancha and Frenz 2000). Additionally, available tools were utilized to make approximations of total arsenic intake from food, water, and air (Table 4). Among arsenic exposure data used were surveys of schoolchildren and adolescents (Ivanovic et al. 1986, 1987), and model diet data on adults recommended by the Ministry of Health. Although measures of consumption were indirect, the tools and surveys used represent good estimates of Chilean food consumption founded upon national cultural eating patterns. Rates used to estimate water and air

**Table 4** Total arsenic exposure by geographic zone among Chilean population groups

| Zone | Total arsenic exposure (ng person $d^{-1}$) | | |
|---|---|---|---|
| | Adults | Adolescents | Preadolescents |
| Northern | 85.1 | 52.8 | 60.9 |
| Central | 42.2 | 29.0 | 23.3 |
| Southern | 23.2 | 17.7 | 12.9 |

intake were also those established by the Chilean Ministry of Health. Once estimates for air, water, and food intake were assembled, calculations of human exposure were made using arsenic concentrations found in these substrates at different points in the country.

Results reveal that, when compared to international standards, individuals living in Chile's northern zone have arsenic exposures that are abnormally high and unhealthy. Results also show a clear tendency of decreasing total arsenic exposure as one progresses from the north to the south of Chile; this decline in exposure appears to closely correlate with the abundance of naturally occurring elements (arsenic, etc.) and the proximity to anthropogenic mining activities.

The relative proportion of total arsenic exposure to adults accounted for by air, food, and water is presented in Table 5. Drinking water constitutes the most important exposure pathway for urban populations in both northern and central Chile. In the south, exposure is similar to that in other countries that do not have the Chilean combination of natural geological arsenic pollution coupled with anthropogenic sources of arsenic (Dabeka et al. 1993). Results show that, nationwide, airborne arsenic is the least significant exposure source for people. However, the situation can be very different at or near mining sites, where this contamination is a relevant health issue. Finally, in contrast to the pattern in Chile, foodstuffs typically represent a high percentage of total arsenic exposure in most countries. This distribution of arsenic exposure by source (air, water, food) correlates closely with the nature of Chilean geology and mining activities, and the fact that Chilean urban populations are supplied with foodstuffs from agricultural and livestock breeding areas that are generally not contaminated by arsenic.

**Table 5** Percent contribution of total arsenic exposure by intake pathway for residents of various Chilean cities

| Zone | Cities | Arsenic contribution (%) | | |
|---|---|---|---|---|
| | | Water | Air | Foods |
| Northern | Arica | 56.91 | 1.58 | 41.51 |
| | Iquique | 89.15 | 0.4 | 10.45 |
| | Antofagasta | 81.77 | 1.46 | 16.78 |
| | Calama | 82.91 | 2.77 | 14.32 |
| | Copiapó | 33.21 | 12.29 | 54.5 |
| | Coquimbo | 46.82 | 1.95 | 51.23 |
| Central | San Felipe | 25.3 | 1.3 | 73.4 |
| | Santiago | 72.7 | 0.9 | 26.4 |
| | Rancagua | 69.3 | 2 | 28.7 |
| | Talca | 46.2 | 0.3 | 53.5 |
| Southern | Concepción | 19.2 | 1.3 | 79.5 |
| | Temuco | 19.4 | 0.4 | 80.2 |
| | Coyhaique | 19.4 | 0.4 | 80.2 |
| | Puntas Arenas | 19.4 | 0.4 | 80.2 |

The evolution of arsenic exposure in Antofagasta was evaluated under the assumption that levels in air and foodstuffs have remained nearly constant whereas those in water have varied. The evaluation indicated that, until the 1980s, water contributed more than 90% of total arsenic exposure; since 2004, that figure declined so that approximately 50% of total exposure comes from water. Simultaneously, the percentage contribution of foodstuffs to arsenic exposure has increased. Before 1990, air represented less than 1% of total arsenic exposure in Antofagasta, whereas since 2004 its contribution to overall arsenic exposure has increased to about 3.3% (Table 6).

## 2.3 Health Effects

High exposure in Chile's north to arsenic during 1950–1970 produced early health effects, including increased rates of infant mortality. Table 7 provides a historic record of average arsenic concentrations in Antofagasta drinking water.

The excess incidence of death for the period 1958–1965 was estimated to be between 18% and 24% (Hopenhayn-Rich et al. 2000). Children were reported to suffer from arsenicosis (Borgoño and Greiber 1971; Bruning 1968) and from arsenicosis associated with respiratory symptoms, including diffuse and segmentary bronchiectasis (Borgoño and Greiber 1971). In the general population, excessive rates of respiratory symptoms and vascular disorders associated with arsenic exposure appeared (Borgoño et al. 1977; Zaldivar 1980; Puga et al. 1973; Rosenberg 1974).

**Table 6** Historic relative contributions of arsenic exposure from water, air, and food among the population of Antofagasta

| Period | Arsenic contribution (%) | | |
|---|---|---|---|
| | Water | Air | Food |
| 1930–1957 | 92.47 | 0.59 | 6.94 |
| 1958–1970 | 99.16 | 0.07 | 0.78 |
| 1971–1979 | 93.76 | 0.49 | 5.80 |
| 1981–1987 | 90.53 | 0.74 | 8.73 |
| 1988–2003 | 84.53 | 1.20 | 14.26 |
| 2004–2006 | 57.73 | 3.29 | 38.97 |

**Table 7** Average concentrations of arsenic in the drinking water of Antofagasta (northern Chile) for various periods since 1930

| Period | Arsenic (µg/L) |
|---|---|
| 1930–1957 | 90 |
| 1958–1970 | 860 |
| 1971–1979 | 110 |
| 1980–1987 | 70 |
| 1988–2003 | 40 |
| 2004–2007 | 10 |

The high risk of death from lung cancer in Antofagasta was first reported and described for the period 1976–1978. Standard Mortality Rates (SMRs) from lung cancer in Antofagasta were five times greater than was the national average (Haynes 1983).

An epidemiological study conducted in 1994–1996 generated data on arsenic exposure and death rates from cancers of the lung, bladder, kidney, and skin for the period 1950–1996. The study authors concluded that the most significant public health effect of the presence of arsenic in drinking water was lung cancer (Ferreccio et al. 2000). The main causes of relative excess of mortality, adjusted for age and sex, in Antofagasta during 1985–1992 were from bladder cancer, followed by lung cancer. The high risk for lung cancer in Antofagasta persisted for 20–30 yr after operation of water treatment plants, built to remove arsenic, began. From 1993 to 2002, the risk of dying from bladder or lung cancer in Antofagasta was four and seven times higher, respectively, than in the rest of Chile (Ferreccio and Sancha 2006). Other epidemiological investigations produced similar results (Smith et al. 1998; Rivara et al. 1997).

These rates are expected to decrease, in the coming years, because of significant diminution of arsenic concentrations in drinking water. However, the impact of increased arsenic emissions into the atmosphere—mitigated somewhat by the trapping of gases in recent years—may have negative consequences in the coming decades.

## 3 Removing Arsenic from Water

### 3.1 Arsenic Removal Technologies Used in Chile

High arsenic water levels exist in the north of Chile, a region characterized by water scarcity. For this reason, water treatment must preserve as much usable water as possible. The literature discloses that the most commonly available technologies for removing arsenic from water are founded on one or more of the following principles: coagulation/precipitation, adsorption, ionic exchange, or membrane filtration. The first, coagulation and precipitation, is a classic process for treating water to remove arsenic; the others are advanced processes of more recent development. Some of the latter processes have the disadvantage of consuming large amounts of water.

Removal of arsenic from water can be a very expensive process. Costs may exceed the means of those benefited, requiring careful selection of the technology to be utilized. The selection must first consider the characteristics of the water to be treated. Characteristics of surface waters and groundwater may be similar or different (Table 8). Some waters contain ions that compete with arsenic for adsorption sites or ionic interchange and thereby may reduce the efficiency of the removal process (Holm 2002; Meng et al. 2000; Clifford 1999).

In Chile, studies undertaken in the 1960s made it possible to appreciate the potential of the coagulation/precipitation process for effectively removing arsenic from water (Latorre 1966). This process was widely known and used at the time, and is still used today. A main feature of the process is that it allows suspended solids removal

**Table 8** Principal characteristics of water found in northern Chile

| Parameter | Unit | Surface water Range | Groundwater Range |
|---|---|---|---|
| pH | | 8.0–8.4 | 7.0–8.0 |
| Total dissolved solids | mg/L | 700–800 | 730–790 |
| Arsenic | ug/L | 400–600 | 60–80 |
| Sulfate | mg/L | 80–100 | — |
| Chloride | mg/L | 120–140 | — |
| Alkalinity | mg/L CaCO3 | 100–120 | 50–60 |
| Hardness | mg/L CaCO3 | 130–150 | 350–400 |
| Silica | mg/L SiO2 | 20–30 | 20–30 |
| Boron | mg/L | 3–4 | 2–5 |
| Dissolved organic carbon | mg/L | Negligible | Negligible |
| Dissolved arsenic | µg/L | 400–600 | 60–80 |
| Particulated arsenic | µg/L | Negligible | Negligible |

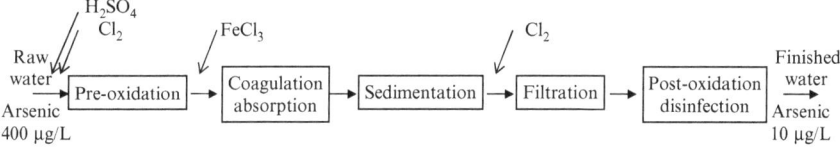

**Fig. 4** General schematic for the most common Chilean arsenic-removal process for surface water

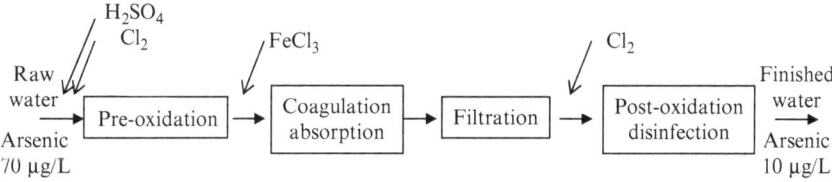

**Fig. 5** General schematic for the most common Chilean arsenic-removal process for groundwater

from water by forming aluminum or iron oxyhydroxides. Research demonstrated that arsenic is adsorbed by such oxyhydroxides, resulting in removal of arsenic in its particulate form. The coagulation/precipitation process removes only arsenic.

Beginning in the 1970s, coagulation was used in Chile for large-scale arsenic removal from both surface waters and groundwater (Figs. 4, 5). In 1990, there were four treatment plants for removing arsenic from surface drinking water sources in Antofagasta and Calama, with a total combined normal treatment capacity of 1,730 L/sec. In 1998, a small plant of 32 L/sec was placed in operation in Taltal to treat groundwater (Table 9).

The efficiency of coagulation in removing arsenic from water depends on such factors as pH of the waters treated, dose of coagulant, and filtrate flux and backwash interval. These factors can all be controlled and adjusted, as required, in the operation of treatment plants. Other factors, such as mixing times and energy usage levels,

**Table 9** Water treatment facilities for arsenic removal in Chile

| Utility | Capacity (L/sec) | Water sources[a] | As range (µg/L) |
|---|---|---|---|
| Salar del Carmen Complex[b] | | | |
| Old plant (1970) | 500 | Toconce | 600–900 |
| | | Lequena | 150–350 |
| New plant (1978) | 520 | Quinchamale | 100–250 |
| | | Siloli Polapi | <50 |
| Cerro Topater[b] (1978) | 500 | Toconce | 600–900 |
| | | Lequena | 150–350 |
| | | Quinchamale | 140–250 |
| Chuquicamata[b] (1989) | 210 | Colana | 70–90 |
| | | Inacaliri | 80–90 |
| Taltal[c] (1998) | 32 | Agua Verde | 60–80 |

[a] The names in this column for surface water sources refer to rivers.
[b] Surface water.
[c] Groundwater.

must be considered in the basic design of the plant. Numerous investigators have studied this first factor (McNeill and Edwards 1995, 1997; Scott et al. 1995; Cheng et al. 1994; Edwards 1994). For treatment purposes, it is necessary to know the As(III)/(V) speciation ratio in the water. As(III), which is a uncharged molecule and a weak ligand, is much more difficult to remove from water than are the anionic As(V) forms, which are readily removed by adsorption onto aluminum or iron oxyhydroxides formed during coagulation processes (Clifford 1999).

Recent studies conducted in Chile, using waters of low turbidity, compared efficiencies and costs of arsenic removal using coagulation and adsorption onto different media. Results show that, in Chile, coagulation is the more cost-effective technology (Fuentealba 2003). In the early years after 2000, the coagulation process was improved to further reduce amounts of arsenic from water in all Chilean water treatment plants (Granada et al. 2003). For Chile, coagulation is a viable technology for meeting the arsenic Maximum Contaminant Level (MCL) of 10 µg/L.

An alternative process, adsorption, has been judged to be disadvantageous for Chile because of both the volume of water consumed during treatment and the high cost of adsorbent material. The limited information available on regenerating absorbent media, and media disposal problems, are additional factors that detract from use of adsorption processes.

Disposal of residuals generated by arsenic removal processes is another important issue and consideration. In Chile, these arsenic residuals were disposed of in the desert during the early years without taking special precautions. Beginning in approximately 2000, disposal began at specially engineered sites. These sites use a geotextile barrier lining beneath disposed arsenic waste, and are then covered with a capping barrier system (Cerda et al. 1999).

## 3.2 Costs

As previously discussed, there are currently four large plants in Chile that treat raw water to remove arsenic for the purpose of producing safe drinking water. The operation of these plants requires the addition of an oxidant and coagulant (chlorine and ferric chloride, respectively) to produce clean water. These plants have successfully met drinking water standards set by different regulations from the 1970s (120 µg/L) to the present (10 µg/L).

In 1997, a goal of the FONDEF project was to evaluate the costs of further reducing arsenic concentrations in drinking water. Table 10 summarizes options developed to effect such reductions for arsenic in drinking water in Antofagasta (Sancha et al. 2000). When judging feasibility, we considered (i) the technology currently used in Chile and its availability on the international market; (ii) the effectiveness of arsenic removal with each technology; (iii) the special problem of water scarcity in northern Chile; and (iv) the coastal location of Antofagasta.

Experience in Chile with coagulation to remove arsenic, along with the results of a new study by González (1997), indicated that coagulation could achieve a residual arsenic level of <0.030 mg/L, if an improved process were used. Technical experience suggested that, to attain a residual level of <0.020 mg/L, a double filtration system would have to be used. Residual arsenic values of 0.010 mg/L or lower could only be achieved using membrane filtration such as reverse osmosis, which requires water preconditioning by coagulation. Another alternative for attaining greater levels of arsenic removal is to mix the effluent from the current treatment system with desalinated seawater.

Figure 6 presents annualized costs of removing arsenic from drinking water in Antofagasta, as evaluated in 1997. These costs include (i) all capital investments needed to implement the removal operation or related costs; (ii) fixed and variable operating costs per year; (iii) repair and maintenance costs; and (iv) any economic credits received as a result of increased water availability, when applying a given option.[1]

**Table 10** Feasible options for removing arsenic from drinking water in Antofagasta

| Options | Targets (mg arsenic/L) | | | | |
|---|---|---|---|---|---|
| | 0.03 | 0.02 | 0.01 | 0.005 | 0.002 |
| Improvement of current system | X | X | | | |
| Reverse osmosis | X | X | X | X | X |
| Desalination | | | X | X | X |

*Source*: FONDEF 2-24 Project.

---

[1] To estimate the cost of each option multiple data sources were used. The assumptions and considerations used in analyzing the different alternatives correspond mainly to information from the regional water and sanitation enterprise (ESSAN). Budgets and input costs were requested from private firms.

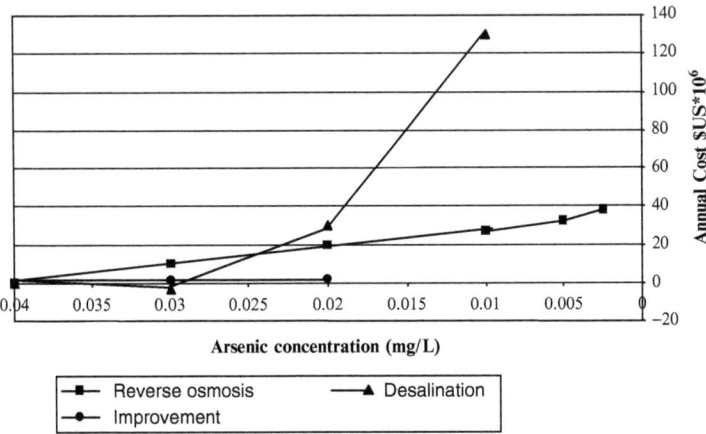

**Fig. 6** Annual cost to treat arsenic in drinking water from Antofagasta

Results show that in, Antofagasta, there were three alternatives, each with its associated costs, to reduce arsenic to 0.03 mg/L from its prevailing 1997 level of 0.04 mg/L. Simple plant process improvement related to the automation of processes produced an annualized total profit increase to the Antofagasta treatment plant of $US 22,000; by incorporating a mixing step with desalinized seawater, there were annualized benefits of $US 5 million; reverse osmosis gives an additional annualized cost of $US 10 million. Savings were achieved from credits arising from an aggregate saving on chemical treatment products; with desalination, greater volumes of freshwater were generated, which negated the need for new investments ($US 170 million) envisioned by the ESSAN (regional water sanitation enterprise) water company to acquire new water sources. On the other hand, no credit was produced with reverse osmosis, which had only costs relating to technology investment and construction.

Consequently, to reduce arsenic to 0.03 mg/L in Antofagasta, it was best to make process improvements or to include the desalination step. If capital investment funds are unavailable, the first option would be most suitable; otherwise desalination is the best approach. If the aim is to reduce the arsenic content in treated water to 0.02 mg/L, double filtration was the option recommended to achieve it. However, using the equipment installed, desalination produces about 1300 L/sec of water in excess of the population's demand. Therefore, it would be attractive if the high cost of installing desalination could be offset by selling excess water volumes generated. The mining sector was a candidate to acquire such water, because they faced a deficit in Chile's north of up to 3000 L/sec for the year 2000. Since then, a small desalinization plant has been built, and currently an expansion is being undertaken to satisfy the increasing demand from mining activity.

The best known option during the 1990s to reduce arsenic levels in treated water to 0.01 mg/L was reverse osmosis. However, to achieve this lower concentration consumes an amount of water equal to an additional 300 L/sec, which in this part

**Table 11** Effect of arsenic level reduction on tariff rates[a]: Antofagasta

| Treatment alternative | Target concentration (mg/L) | Rate increase ($US/m³) |
|---|---|---|
| Inverse osmosis | 0.03 | 0.27 |
| | 0.02 | 0.53 |
| | 0.01 | 0.68 |
| | 0.005 | 0.79 |
| | 0.002 | 0.83 |
| Desalination | 0.03 | −0.13 |
| | 0.02 | 0.45 |
| | 0.01 | 1.03 |

*Source*: FONDEF 2–24 Project.
[a] Tariff rates refer to the periodic charges made to households to cover the cost of treated water.

of the country, where water is scarce, has a very high value. The higher water consumption renders this alternative unattractive from an economic perspective.

In Chile, water tariffs (or costs) are paid at the household level and are based on the costs of the water provided to the user. Consequently, the costs of the different options discussed above had a potential impact on water tariffs. The effect on tariffs is depicted in Table 11. The price of water in the north of Chile was $US 1.21/m³ in 1997.[2] Desalination allows tariff rates to be lowered by $US 0.13/m³ of water (equivalent to 11% of the price), if a quality of 0.03 mg/L is required. More stringent requirements (lower arsenic concentration) translates as significant increases in rates of between $US 0.5 and $US 1/m³, equivalent to a premium of 41% and 83%, respectively, over current water prices in Chile's north.

In conclusion, results show that residual arsenic in Chile's drinking water could be substantially reduced in the late 1990s; however, costs increased significantly as arsenic removal rates increased, particularly when arsenic concentrations approached 0.01 mg/L. For this reason, evaluation of lower cost options for Antofagasta continued. In recent years, optimization of the coagulant process has resulted in achieving considerably lower arsenic concentrations in treated water (0.01 mg/L) at a relatively lower cost (Granada et al. 2003). Consequently, beginning in 2006, the more stringent standard of 0.01 mg/L was imposed, and it is now complied with in all treatment plants in the north of Chile.

## 4 Reducing Airborne Arsenic Emissions

In 1995, nearly 90% of Chile's copper was produced by seven smelters using pyrometallurgy. Air concentrations of arsenic in the vicinity of smelters were particularly high. The high levels derive both from high arsenic content in Chile's copper ore, typically reaching 1% (more or less the same as the copper content), and the fact that the smelting process did not incorporate antipolluting technologies.

---

[2] (Source: ESSAN).

Consequently, in 1995, air concentrations of arsenic reached average annual values of about 1 μg/m$^3$ in the area of many smelters. In the worst case, average annual values in air reached 10 μg/m$^3$, with daily maximums of 100 μg/m$^3$. Total emissions from smelters in 1992 reached almost 9,000 t/yr.

## 4.1 Control Options

Two primary approaches are generally used to reduce arsenic air emissions. The first employs technology aimed at reducing both chimney stack and fugitive emissions. The aim of the second approach is to reduce the exposure of people to arsenic air levels. Technology improvements are designed to modernize existing processes and implement systems for capturing and treating effluent gases. Such technological changes permit better control of gas emissions. If capture and treatment of metallurgical gases is insufficient in any one plant, more profound technological measures may be used, including replacement of older smelters with more modern technology.

The second approach, designed to reduce arsenic exposure to the population, includes changes in the arsenic content of ores fed into the smelter, reduction in daily activity levels for people in arsenic-polluted areas, and even relocation of local populations, if necessary. Indeed, it is possible to reduce arsenic emissions by feeding the smelter with a mix of copper ores with different arsenic content, pre-roasting of concentrated feed, or by reducing the quantity of concentrate fed into the process. Finally, by relocating the affected population it is possible to reduce air arsenic exposure without modifying smelter operations.

Table 12 summarizes ranges of air emission reduction achievable by implementing each of the foregoing options.

**Table 12** Summary of reduction efficiencies for various As reduction approaches

| Option | Capture efficiency (%) |
|---|---|
| Electrostatic precipitators | 10–50 |
| Acid plants | +99 |
| Flash Inco Oven | 95 |
| Contop technology | 95 |
| Isasmelt Process | 95 |
| Flash Converting-Flash Smelting | +99 |
| Mitsubishi | +99 |
| Multi-story roaster | 90 |
| Dust treatment plant | 80–90 |
| Secondary hood | 94–99 |
| Matte and slag launder cover | 99 |
| Smelter shutdown | 100 |
| Change of concentrate feed | Variable |
| Reduction of activity | Variable |
| Population relocation | Variable |

*Source*: FONDEF 2–24 Project.

## 4.2 Costs to Reduce Arsenic Air Pollution

Reducing pollution always has associated costs and benefits, and both must be considered when choosing and implementing programs to achieve lower exposures. Of course, there is a trade-off between costs and degree of reduced arsenic exposure that is achievable: the lower the cost input, the higher the exposure. To inform policymakers of their options, a detailed cost analysis was conducted at different levels of arsenic emission reduction. This analysis allowed decision makers to choose the most cost-effective approach that would offer protection to exposed populations at levels considered to be acceptable.

Costs of options include capital investment expenditures, as well as the operating and maintenance costs, with possible benefits, such as credits arising from generation of additional sulfuric acid, being factored in. Costs were calculated for each smelter based on literature information, data from national and international experts, and data from smelter operations. Indirect costs such as job loss, reduction of ancillary services, etc., were not considered.

Annualized costs were determined by calculating the comparative net present value (NPV) of the existing smelter operation with and without the emission reduction project. An investment horizon of 25 yr with a 12% discount rate was used in the NPV calculation. Table 13 presents a typical smelter cost profile. Similar data

**Table 13** Cost of reducing emissions with technology and by relocation of population: an example showing private costs and incomes before tax

| Option | Description of option | Emissions (t/d) | Maximum concentration ($\mu g/m^3$) | Incremental investment ($US millions) | Net cost (PV) ($US millions) | Net cost/ annum ($US thousands) | Net cost of anodes (US cents per lb) |
|---|---|---|---|---|---|---|---|
| 0 | Base case | 1.4 | 6.01 | 0 | 0 | 0 | 0 |
| 1 | Improved base case | 0.43 | 1.4 | 5.1 | 4.1 | 518 | 0.2 |
| 2 | Teniente converter | 0.65 | 2.1 | 72.9 | −29.6 | −3,773 | −1.1 |
| 3 | Contop | 0.21 | 0.82 | 54.7 | −100.1 | −12,773 | −3.6 |
| 4 | Flash furnace | 0.74 | 2.14 | 159.1 | 103 | 13,139 | 3.7 |
| 5 | Continuous smelting | 0.08 | 0.46 | 352 | 150.7 | 19,200 | 5.4 |
| 6 | Smelter shutdown | 0 | 0.08 | 0 | 49.9 | 6,361 | 2.3 |
| 7 | Relocation of population | 1.4 | 0.08 | – | 30 | 3,800 | 1.4 |

PV, present value.
*Source*: FONDEF 2–24 Project.

were prepared for each smelter.[3] The smelter depicted in Table 13 had two dominant options. In the first option, the plant could switch to Contop Technology with net benefits of $US 13 million, a result of net credits from selling sulfuric acid, a production by-product. Emissions of arsenic would be reduced to approximately one-seventh of the original value. However, such levels are still rather high. So, if additional reduction is required, the best option for this smelter is to relocate the affected population (workers and their families), comprising 3,000 people. This latter alternative was the one finally adopted; this relocation was completed in 2007.

## 5 Setting Environmental Standards for Arsenic: Present and Future

### 5.1 Air

Until 1990, there was no regulation in Chile for airborne arsenic emissions. In that year, the new democratic government declared that environmental issues would be a main concern for the administration (O'Ryan and Lagos 2005). One of the priority environmental problems facing the new administration was that of arsenic emissions from copper and gold smelters. In particular, the Health Ministry was pushing for a strict ambient emissions standard. A special Interministerial Commission for Air Quality (CICA) was established with participation from four ministries: Agriculture, Health, Economics, and Mining (Camus and Hajek 1998). A deliberative group was formed that included public and private institutions, and also included private mining firms and the State-owned National Copper Corporation (CODELCO).

Based on available information, a first regulation was established at the end of 1991 through "Decreto 185." This regulation imposed an emission standard on all sources of emitted arsenic. It was recognized that, before developing more elaborate emission standards, a great deal of additional information would be required. Such information included emissions and dispersion data, health effects, control options, and associated costs. The FONDEF project was proposed, and two relevant regulatory questions were posed: (1) What is the appropriate regulatory approach: an ambient standard or emissions standard for each source? and (2) How strict should the standards be? The first question was hotly debated, because an ambient standard was preferred by many institutions who were most concerned about the high arsenic concentrations observed in the north of Chile. Others, however, were concerned about the high economic costs and social consequences of imposing a uniform ambient standard[4] that would universally apply, even in localities with very low population exposure. A quantitative evaluation was made to answer open

---

[3] For detailed results see O'Ryan and Diaz (2000).
[4] Ambient standards aimed at protecting health must be uniform in the whole country.

**Table 14** Permissible arsenic emission air standards in Chile

| Location or Province | Capacity (t/yr) | Arsenic emission allowed (t/yr) | | |
|---|---|---|---|---|
| | | 2000 | 2001 | 2003 |
| El Loa | ≤1,400,000 | 1,100 | 800 | 400 |
| Antofagasta | ≤350,000 | 126 | | |
| Copiapó | ≤200,000 | 42 | | 34 |
| Chañaral | ≤500,000 | 1,450 | 800 | 150 |
| Del Elqui | ≤80,000 | 200 | | |
| San Felipe de A. | ≤350,000 | 95 | | |
| Valparaíso | ≤400,000 | 120 | | |
| Cachapoal | ≤1,100,000 | 1,880 | 375 | |

questions by studying the main profile of each smelter, including costs, emissions, and arsenic concentrations at the most affected locations (O'Ryan and Díaz 2000).

Results showed that a unique ambient standard could not be too stringent without jeopardizing smelter operations; in fact, stringent standards would result in a shutdown of operations for some smelters. However, it was also recognized that lax ambient standards would encourage plants that had high emissions (and their future investors) to wrongly believe that it would be permissible to build plants with relatively low emission controls. Consequently, it was decided to regulate using emission standards specific to each region in which pollution sources where located. Using a reference ambient quality standard for local residents (between 0.1 and 0.2 µg/m$^3$), regulators imposed the emission standards presented in Table 14. This regulation was passed on June 2, 1999, as Decree 165/1999 MINSEGPRES (Ministerio Secretaria General de la Presidencia, or in English, Ministry General Secretariat of the Presidency), and it is currently in force.

## 5.2 Water

Regulating permissible levels of arsenic in potable water has been a constant concern in Chile (Table 15). Given Chile's special geological conditions, adopting the MCL recommended by the WHO has been impossible, both technically and economically. Instead, over a period of four decades, Chile has established a series of ever-lower feasible MCLs to improve standards for arsenic in drinking water, and these standards have continued to be modified downward as scientific studies and techniques for optimizing removal processes have progressed.

Thus, in 1970, the Chilean standard for permissible arsenic in drinking water was set at 0.12 mg/L, because that was the level of arsenic removal achieved by the first treatment plant built in Chile. Gradually, after Chilean researchers determined how to reduce residual levels of arsenic to 0.05 mg/L by means of small modifications in plant design and operation, this value became the legal limit in 1984. Finally, at the beginning of 2000, the reliability of the optimized coagulant process permitted reaching 0.01 mg/L, allowing imposition of this level as the standard.

**Table 15** Evolution of Chilean regulations governing arsenic content of drinking water

| Year | Standard | MCL (mg/L As) |
|------|----------|---------------|
| 1970 | NCh 409Of70 Drinking water: Requirements | 0.12 |
| 1984 | NCh 409Of84 Drinking water: Part 1: Requirements | 0.05 |
| 2005 | NCh 409/1Of2005 Drinking water: Part 1: Requirements | 0.01 |

MCL, Maximum Contaminant Level.

Arsenic of no more than 0.01 mg/L in potable water became established as the Chilean standard in 2005. This standard governs practices of arsenic removal in plants constructed since 2005. Plants constructed before to 2005 have adopted timetables for eventual compliance with the new norm, and indeed some of these reached the new standard by 2004. Those few plants constructed before 2005 that have not yet achieved the norm of 0.01 mg/L have a 5-yr window in which to achieve a residual arsenic level of 0.03 mg/L and a maximum of 10 yr to reach 0.01 mg/L.

# 6 Concluding Remarks

Regulating hazardous pollutants requires a careful weighing of environmental and economic trade-offs and, often, gradualism in implementing needed changes. Research fitted to the context of developing countries is particularly important, because copying standards from other countries may neither be affordable nor necessarily produce appropriate risk reductions. In the context of developing countries, it is essential that compliance be affordable and feasible.

It is important to prioritize environmental risks and to focus on the most important ones first. Following this concept, ex post studies in Chile show that it was appropriate (for arsenic) that priority first be given to regulation of water and subsequently of air. Of course, it would be particularly convenient that riskcost analysis be performed *ex ante* to most effectively allocate resources.

Arsenic regulation in Chilean water was preceded by development of appropriate technologies to adequately address and reduce arsenic content. This effort began over 40 yr ago, when this pollutant was first identified as a health hazard. Average concentrations of arsenic in drinking water have fallen from 860 µg/L, more than four decades ago, to 110 µg/L after 1971; currently, average concentrations are ~10 µg/L.

In air, the regulatory process has developed much more recently and has been centered on copper smelters. In regulating air emissions of arsenic for application to copper smelters, technical feasibility costs and risks of proposed standards were all weighed. As a result, different emission standards were applied to each smelter, and application of a universal and uniform ambient standard was discarded until more definitive information becomes available.

The regulatory process and associated research needed to create regulations has been a joint effort among university researchers, regulatory agencies, and entities in the private sector, particularly the mining industry. This collaboration has allowed a fruitful interchange of information and experience and has enhanced the regulatory process. However, much progress is yet to be made, in particular, to meet the established arsenic standard for all Chilean water sources and also to make progress in achieving goals for reductions in airborne arsenic concentrations.

## 7 Summary

Chile is one of the few countries that faces the environmental challenge posed by extensive arsenic pollution, which exists in the northern part of the country. Chile has worked through various options to appropriately address the environmental challenge of arsenic pollution of water and air. Because of cost and other reasons, copying standards used elsewhere in the world was not an option for Chile.

Approximately 1.8 million people, representing about 12% of the total population of the country, live in arsenic-contaminated areas. In these regions, air, water, and soil are contaminated with arsenic from both natural and anthropogenic sources. For long periods, water consumed by the population contained arsenic levels that exceeded values recommended by the World Health Organization. Exposure to airborne arsenic also occurred near several large cities, as a consequence of both natural contamination and the intensive mining activity carried out in those areas. In rural areas, indigenous populations, who lack access to treated water, were also exposed to arsenic by consuming foods grown locally in arsenic-contaminated soils.

Health effects in children and adults from arsenic exposure first appeared in the 1950s. Such effects included vascular, respiratory, and skin lesions from intake of high arsenic levels in drinking water. Methods to remove arsenic from water were evaluated, developed, and implemented that allowed significant reductions in exposure at a relatively low cost. Construction and operation of treatment plants to remove arsenic from water first began in the 1970s.

Beginning in the 1990s, epidemiological studies showed that the rate of lung and bladder cancer in the arsenic-polluted area was considerably higher than mean cancer rates for the country. Cancer incidence was directly related to arsenic exposure.

During the 1990s, international pressure and concern by Chile's Health Ministry prompted action to regulate arsenic emissions from copper smelters. A process began in which emission standards appropriate for Chile were set; this process included careful evaluation of risks versus mitigation costs for abatement options. Such options were developed and implemented.

More recently, local communities have pressed for more significant reductions of arsenic in air and water. Considerable experience was gained with the arsenic experience on how to manage this type of hazardous pollutant, in a context of trade-offs among production, jobs, income, and health. In this review article, we cover arsenic levels in Chile's air, water, and soils and discuss health impacts and patterns of exposure. We also describe the process followed to set arsenic regulatory standards, as well as abatement options for air and water and the associated costs.

**Acknowledgments** The authors wish to thank all those who made this study possible. We are grateful to the mining and sanitary companies that participated in the project and provided valuable technical support: in particular, Corporación Nacional del Cobre de Chile (CODELCO), Compañía Minera Disputada de las Condes, Compañia Minera El Indio, Empresa Nacional de Mineria (ENAMI), and REFIMET S.A. We also express our gratitude to the numerous students and technical staff who, with their hard work and enthusiasm, made this study possible. We are especially grateful to the international support of experts from the U.S. Environmental Protection Agency (EPA), Carnegie-Mellon, and World Health Organization (WHO). We acknowledge the financial support of Fund for the Promotion of Scientific and Technological Development (FONDEF). Finally, we thank the University of Chile for its support on this project.

# References

Alam M, Snow T, Tanaka A (2003) Arsenic and heavy metal contamination of vegetables grown in Samta village, Bangladesh. Sci Total Environ 308:83–96.
Ayers RS, Westcot DW (1994) Water quality for agriculture. Irrigation and Drainage Paper 29, rev. 1. Reprinted 1989, 1994. Food and Agriculture Organization of the United Nations, Rome (1985).
Borgoño J, Greiber R (1971) Epidemiologic study of arsenic poisoning in the city of Antofagasta (Estudio Epidemiológico del arsenicismo en la ciudad de Antofagasta). Rev Méd Chil 99:702–707.
Borgoño J, Vicent O, Venturino H, Infante A (1977) Arsenic in the drinking water of the city of Antofagasta: epidemiological and clinical study before and after the installation of a treatment plant. Environ Health Perspect 19:103–105.
Bruning W (1968) El problema del hidroarsenicismo crónico regional endémico en Antofagasta. Rev Chile Pediatr 39:49–51.
Camus P, Hajek E (1998) Transición a la democracia y medio ambiente 1990 – 1994. Historia ambiental de Chile. Doc electr: http://www.hajek.cl/ecolyma/doc03d.htm.
Cerda W, Gatica R, Veneros M (1999) Sistema de recuperación de aguas de descarte y disposición final de lodos arsenicados. Rev AIDIS-Chile 24:26–31.
Clifford DA (1999) Ion Exchange and Inorganic Adsorption, 5th Ed. Water Treatment and Disposal. McGraw-Hill, New York.
Cheng R, Liang S, Wang H, Buehler M (1994) Enhanced coagulation for arsenic removal. J Am Water Works Assoc 86(9):79–90.
Dabeka R, McKenzie A, Lacroix G, Cleroux C, Bowe S, Graham R, Conacher H, Verdier P (1993) Survey of arsenic in total diet food composites and estimation of the dietary intake of arsenic by Canadian adults and children. JAOAC Int 76:14–25.
De Gregori I, Fuentes E, Rojas M, Pinochet H, Potin-Gautier M (2003) Monitoring of copper, arsenic and antimony levels in agricultural soils impacted and non-impacted by mining activities from three regions in Chile. J Environ Monit 2:287–295.
De Gregori I, Fuentes E, Olivares D, Pinochet H (2004) Extractable copper, arsenic and antimony by EDTA solution from agricultural Chilean soils and its transfer to alfalfa plants (*Medicago sativa* L.). J Environ Monit 6:1–11.

Diaz OP, Leyton I, Munoz O, Nunez N, Devesa V, Suner MA, Velez D, Montoro R (2004) Contribution of water, bread, and vegetables (raw and cooked) to dietary intake of inorganic arsenic in a rural village of Northern Chile. J Agric Food Chem 52(6):1773–1779.

Edwards M (1994) Chemistry of Arsenic removal during coagulation and Fe-Mn oxidation. J Am Water Works Assoc 86(9)64–78.

Enriquez H (1978) Relación entre el contenido de Arsénico en Agua y el Volcanismo Cuaternario en Chile, Bolivia y Perú. UNESCO. Documentos Técnicos en Hidrología, Montevideo.

EHC (1981) Environmental Health Criteria, No. 18, Arsenic.

Ferreccio C, Sancha A (2006) Arsenic exposure and its impact on health in Chile. J Health Popul Nutr 24(2):164–175.

Ferreccio C, Gonzalez C, Milosavjlevic V, Marshall G, Sancha A, Smith A (2000). Lung cancer and arsenic concentrations in drinking water in Chile. Epidemiology 11:673–679.

Flynn H, McMahon V, Chong G, Demergasso C, Corbisier F, Mehard A, Paton G (2002) Assessment of bioavailable arsenic and copper in soils and sediment from the Antofagasta region of northern Chile. Sci Total Environ 286:51–59.

FONDEF 2–24 (1997) Proteccion de la Competitividad de los Productos Mineros de Chile: Antecendentes y Criterios para la Regulación Ambiental del Arsénico. Final Report. Universidad de Chile, November, 1997. Project Director Ana Maria Sancha, Assistant Director Raul O'Ryan. Financed by FONDEF Conicyt.

Fuentealba C (2003) Planta piloto para remover arsénico en una fuente subterránea de agua potable. Dissertation. Facultad de Ciencias Físicas y Matemáticas, Universidad de Chile.

Gonzalez K (1997) Alternativas de remoción de arsénico desde fuentes de agua potable en la II región Antofagasta y costos asociados. Dissertation. Facultad de Ciencias Físicas y Matemáticas, Universidad de Chile.

Granada J, Cerda W, Godoy D (2003) ESSAN: el camino para reducir notoriamente el arsénico en el agua potable. Rev AIDIS-Chile 34:44–49.

Haynes R (1983) The geographical distribution of mortality by cause in Chile. Soc Sci Med 17:355–364.

Holm T (2002) Effects of $CO_3^{-2}$/bicarbonate, Si and $PO_4^{-3}$ on arsenic sorption to HFO. J Am Water Works Assoc 94:174–181.

Hopenhayn-Rich C, Browning S, Hertz-Picciotto I, Ferreccio C, Peralta C, Gibb H (2000) Chronic arsenic exposure and risk of infant mortality in two areas of Chile. Environ Health Perspect 108(7):667–673.

Ivanovic D, Aguayo M, Vásquez M, Truffelh I, Ballester D, Zacarias I (1986) Ingesta dietaria de escolares que egresan de educación básica en el Area Metropolitana de Santiago, Chile. Arch Latinoam Nutr 36:379–400.

Ivanovic D, Zacarias I, Vásquez M (1987) Ingesta dietaria de escolares adolescentes que egresan de educación media en el Area Metropolitana de Santiago de Chile. Rev Med Chile 115:1029–1038.

Latorre C (1966) Estudio físico-químico para la remoción del arsénico en el río Toconce. Revista Ingeniería Sanitaria 12–21, Santiago, Chile.

McNeill LS, Edwards M (1995) Soluble arsenic removal at water treatment plants. J Am Water Works Assoc 87(4):105–113.

McNeill LS, Edwards M (1997) Predicting As removal during metal hydroxide precipitation. J Am Water Works Assoc 89(1):75–86.

Meng X, Bang S, Korfiatis G (2000) Effects of silicate, sulfate and carbonate on arsenic removal by ferric chloride. Water Res 34:1255–1261.

Munoz O, Diaz O, Leyton I, Nunez N, Devese V, Suner M, Velez D, Montoro R (2002) Vegetables collected in the cultivated Andean area of northern Chile: total and inorganic arsenic contents in raw vegetables. J Agric Food Chem 50:642–627.

O'Ryan R, Díaz M (2000) Risk-cost analysis for the regulation of airborne toxic substances in a developing context: the case of arsenic in Chile. Environ Resour Econ 15(2):115–134.

O'Ryan R, Lagos C (2005) Gestión Ambiental Chilena 1990–2005: Avances y Desafíos. In: Meller P (ed) La Paradoja Aparente. Equidad y Eficiencia: Resolviendo el Dilema. Taurus Impresores, Santiago, pp 529–575.

Puga F, Olivos P, Greiber R, Gonzalez I, Heras E, Barrera S, Gonzalez E (1973) Hidroarsenicismo crónico en Antofagasta. Estudio epidemiológico y clínico. Rev Chil Pediatr 44(3):215–222.

Queirolo F, Stegen S, Mondaca J, Cortes R, Rojas R, Contreras C, Munoz L, Schwuger MJ, Ostapczuk P (2000a) Total arsenic, lead, cadmium, copper, and zinc in some salt rivers in the northern Andes of Antofagasta, Chile. Sci Total Environ 255:85–95.

Queirolo F, Stegen S, Restovic M, Paz M, Ostapczuk P, Schwuger MJ, Munoz L (2000b) Total arsenic, lead, and cadmium levels in vegetables cultivated at the Andean villages of northern Chile. Sci Total Environ 255:75–84.

Rivara MI, Cebrian M, Corey G, Hernandez M, Romieu I (1997) Cancer risk in an arsenic-contaminated area of Chile. Toxicol Ind Health 13:321–338.

Rosenberg H (1974) Systemic arterial disease and chronic arsenicism in infants. Arch Pathol 97:360–365.

Roychowdhury T, Uchinot T, Tokunaga H, Ando M (2002) Survey of arsenic in food composites from an arsenic-affected area of West Bengal India. Food Chem Toxicol 40:1611–1621.

Roychowdhury T, Tokunaga H, Ando M (2003) Survey of arsenic and other heavy metals in food composites and drinking water and estimation of dietary intake by the villagers from an arsenic-affected area of West Bengal, India. Sci Total Environ 308:15–35.

Sancha A, Frenz P (2000) Estimate of the current exposure of the urban population of northern Chile to arsenic. In: Reichard E, Hauchman F, Sancha A (eds) Interdisciplinary Perspectives on Drinking Water Risk Assessment and Management. Proceedings of the Santiago (Chile) Symposium). IAHS Publication 260, pp 3–8.

Sancha AM, Vega F, Venturino H, Fuentes S, Salazar AM, Moreno V, Baron AM, Rodriguez D (1992a) The arsenic health problem in northern Chile. Evaluation and control. A case study preliminary report. In: International Seminar Proceedings. Arsenic in the Environment and Its Incidence on Health. Universidad de Chile, Chile, pp 187–202.

Sancha AM, Vega F, Fuentes S (1992b) Speciation of arsenic present in water inflowing to the Salar del Carmen treatment plant in Antofagasta, Chile, and its incidence on the removal process. In: International Seminar Proceedings. Arsenic in the Environment and Its Incidence on Health. Universidad de Chile, Chile, pp 183–186.

Sancha AM, Rodriguez D, Vega F, Fuentes S, Salazar AM, Venturino H, Moreno V, Baron AM (1995) Exposure to arsenic of the Atacameño population in northern Chile. In: Assessing and Managing Health Risks from Drinking Water Contamination: Approaches and Applications. Proceedings of the Rome Symposium. IAHS Publication No. 233, pp 141–146.

Sancha AM, Vega F, Fuentes S, Venturino H, Baron AM, Moreno V, Salazar AM (1997) Exposición a Arsénico de la población atacameña In: 2° Simposio Internacional de Estudios Altiplanitos, Universidad de Chile, pp 109–115.

Sancha A, O'Ryan R, Perez O (2000) The removal of arsenic from drinking water and associated costs: the Chilean case. In: Reichard E, Hauchman F, Sancha A (eds) Interdisciplinary Perspectives on Drinking Water Risk Assessment and Management. Proceedings of the Santiago (Chile) Symposium). IAHS Publication 260, pp 17–25.

Scott K, Green J, Do H, Mc Lean S (1995) Arsenic removal by coagulation. J Am Water Works Assoc 87(4):114–126.

Smith A, Goycolea M, Haque R, Biggs M (1998) Marked increase in bladder and lung cancer mortality in a region of northern Chile due to arsenic in drinking water. Am J Epidemiol 147:660–669.

Ulriksen P, Cabello A (2004) Arsenic concentration in atmospheric particulate material in Chile. In: Sancha A, O'Ryan R (eds) The Environmental Regulation of Toxic Substances: The Case of Arsenic in Chile. Universidad de Chile, Chile, pp 21–33.

Zaldivar R (1980) A morbid condition involving cardio-vascular, bronchopulmonary, digestive and neural lesions in children and young infants after dietary arsenic exposure. Zentbl Bakteriol I Abt Orig B 170:44–56.

# Atrazine Interaction with Estrogen Expression Systems

J. Charles Eldridge, James T. Stevens, and Charles B. Breckenridge

## Contents

1 Introduction ........................................................................................................... 147
2 Effects on the Hypothalamic-Pituitary-Gonadal (HPG) Axis in Rodents ........................ 148
3 Studies on Estrogen Expression Systems .............................................................. 148
    3.1 Estrogen-Mediated Expression *In Vivo* ........................................................... 148
    3.2 *In Vitro* Expression Systems ........................................................................ 152
    3.3 Estrogen Receptor Binding ........................................................................... 154
4 Summary ............................................................................................................. 155
References ............................................................................................................. 156

## 1 Introduction

Atrazine (2-chloro-4-ethylamino-6-isopropylamino-*s*-triazine) is a herbicide used to control growth of many broadleaf weeds and some annual grasses, particularly in corn and sorghum crops (Vencill 2002). Atrazine inhibits plant photosynthesis by binding to the D1 protein of the protein-bound plastoquinone ($Q_b$) located in the plant thylakoid membrane (Devine et al. 1993), thereby preventing electron transfer at the reducing site of chloroplast complex II (Good 1961). Atrazine and its mono- and dideälkylated metabolites are moderately mobile in soil (Qiao et al. 1996), are persistent in the environment, with a soil half-life of 17–26 d (Winkelmann and Kliane 1990), and are detected in surface water (Schottler et al. 1998). Because of these characteristics, the EPA Office of Drinking Water has established an annual average maximum contaminant level (MCL) for atrazine of 3 ppb (US EPA 1996). More recently, the Office of Pesticide Planning has determined that a 90-d average concentration of 12.5 ppb in finished drinking water is safe, based on the chronic no observable effect level (NOEL) of 1.8 mg/kg/d observed in a chronic rodent feeding study (US EPA 2006).

---

J.C. Eldridge, J.T. Stevens
Department of Physiology and Pharmacology, Wake Forest University School of Medicine, Winston-Salem, NC 27157-1083, USA
e-mail: eldridge@wfubmc.edu

C.B. Breckenridge
Human Safety Assessment, Syngenta Crop Protection, Inc.18300, Greensboro, NC 27419-8300, USA

## 2 Effects on the Hypothalamic-Pituitary-Gonadal (HPG) Axis in Rodents

Acute or chronic treatment of female Sprague–Dawley (SD) or Long Evans (LE) rats with atrazine disrupts estrous cycling (Eldridge et al. 1994, 1999a,b; Simpkins et al. 1998; Cooper et al. 2000). Mode of action research has established that this disruption may be attributed to atrazine inhibition of a surge of pituitary luteinizing hormone (LH) necessary for rodent ovulation (Simpkins et al. 1998; Eldridge et al. 1999a; Cooper et al. 2000). Long-term feeding studies of intact SD female rats with atrazine also demonstrated a significantly increased incidence, or earlier lifetime appearance of mammary tumors; such tumors were attributed to prolonged exposure to endogenous estrogen in atrazine-treated animals that developed a state of constant estrus earlier in life (neuroendocrine aging) than did the controls (Eldridge et al. 1994, 1999a; Stevens et al. 1999; Wetzel et al. 1994). The U.S. Environmental Protection Agency (US EPA 2002) and other regulators (IARC 1999; European Union 2000; United Kingdom Pesticide Directorate 2000; APVMA 2004) concluded that the mammary tumor response observed in female SD rats was not relevant to humans because of differences between species in the mechanisms of neuroendocrine aging of the HPG axis.

Because ovarian estrogens mediate neuroendocrine control of pituitary gonadotropin surges, the question arose as to whether the mechanism of atrazine effects on the estrous cycle in female rats might include direct interaction with estrogen expression.

Indeed, some authors suggest that atrazine is "estrogenic" or is an "environmental estrogen" (Davis et al. 1993; Steingraber 1997; Muir et al. 2004; Tanaka et al. 2004), despite the absence of supporting evidence in the literature. Environmental estrogens have long been suspected of etiological involvement in hormone-dependent cancer (Sonnenschein and Soto 1998; Mukherjee et al. 2006), although firm association has been difficult to establish (Calle et al. 2002; Safe 2004; Mitra et al. 2004). The hypothesis that atrazine is estrogenic has been repeatedly evaluated in the published literature over the years. Several dozen studies have addressed whether atrazine can imitate, inhibit, or otherwise modulate estrogen expression, in a wide variety of animal and *in vitro* models. This review addresses and summarizes the results of these studies.

## 3 Studies on Estrogen Expression Systems

### 3.1 *Estrogen-Mediated Expression In Vivo*

Table 1 presents an overview, taken from the literature, of responses to atrazine exposure during testing of estrogen expression *in vivo*.

**Table 1** Responses to atrazine exposure using *in vivo* tests of estrogen expression

| Species | Tissue | Stimulated[a] | Inhibited[a,b,c] | Reference |
|---|---|---|---|---|
| Rat, OVX | Mammary tumor promotion | No | — | Stevens et al. 1999 |
| Rat, OVX | Uterine weight | No | Weak | Tennant et al. 1994a |
| Rat, OVX | Uterine weight | No | Weak | Connor et al. 1996 |
| Rat, OVX | Progesterone receptor expression | No | Weak | Tennant et al. 1994a |
| Rat, OVX | Progesterone receptor expression | No | Weak | Connor et al. 1996 |
| Rat, OVX | Progesterone receptor mRNA | No | No | McMullin et al. 2004 |
| Rat, OVX | Uterine thymidine incorporation | No | Weak | Tennant et al. 1994a |
| Rat, OVX | Uterine peroxidase reaction | No | Weak | Connor et al. 1996 |
| Rat, OVX | Transplanted pituitary tumor | No | — | Fujimoto and Honda 2003 |
| Rat, intact | Reproductive tract development | No | — | Eldridge et al. 1998 |
| Rat, intact | Vaginal cytology cornification | No | — | Eldridge et al. 1999b |
| Rat, intact | Prepubertal uterine weight | No | — | Ashby et al. 2002 |
| Rat, intact | DMBA-induced tumor growth | No | Inconclusive | Tanaka et al. 2004 |
| Rat, OVX | Estrogen-primed LH surge | — | Weak | Cooper et al. 2000 |
| Rat, OVX | Estrogen-primed LH surge | — | Weak | Eldridge et al. 1999a |
| Rat, OVX | Estrogen-primed LH surge | — | Weak | Simpkins et al. 1998 |
| Rat, OVX | Estrogen-primed LH surge | — | Weak | McMullin et al. 2004 |
| Alligator | Eggs, female phenotype | No | — | Crain et al. 1999 |
| Crocodile | Eggs, female phenotype | No | — | Beldomenico et al. 2007 |
| Turtle | Feminization of testicular tissue | No | — | De Solla et al. 2006 |
| Frog | Female sex ratio of tadpoles | No | — | Carr et al. 2003 |
| Goldfish | Vitellogenin production | No | — | Spano et al. 2004 |
| Carp | Vitellogenin production | No | No | Sanderson et al. 2001 |
| Zebrafish | Vitellogenin mRNA | No | — | Muncke et al. 2007 |
| Minnow | Various female reproductive parameters | No | — | Bringolf et al. 2004 |
| Japanese quail | Female oviduct weight, serum LH | No | — | Wilhelms et al. 2006 |
| Japanese quail | Feminization of male reproductive tract | No | No | Wilhelms et al. 2005 |
| Fruit flies | Yolk protein genes | No | No | LeGoff et al. 2006 |

OVX, ovariectomized; —, assessment not done; DMBA, dimethylbenzanthracene; LH, luteinizing hormone.

[a] No, indicates no response to atrazine was observed at doses that were at least $10^5$ molar excess of the effective dose of estradiol.

[b] Weak, indicates a response to atrazine was observed only at doses that were at least $10^5$ in excess of the effective dose of estradiol.

[c] Inconclusive, no dose response or no historical control data for DMBA-treated rats.

A reproductive toxicology study in rats fed atrazine in the diet at concentrations up to 500 ppm gave early evidence that *in vivo* exposure to atrazine was not estrogenic (Hauswirth and Wetzel 1998). Estrogen-dependent responses such as feminization of immature males, premature puberty in females, and infertility were not observed in this study. Additional insights were also obtained from chronic feeding studies. Although several chronic feeding studies with atrazine in intact SD female rats had yielded an increased or earlier lifetime incidence of mammary tumors, the tumor response failed to appear in studies with ovariectomized (OVX) female rats (Stevens et al. 1999). In atrazine-treated OVX rats, well-established estrogen target organs such as mammary and uterine tissues remained rudimentary and appeared unstimulated after long-term feeding at doses that exceeded the Maximum Tolerated Dose (MTD) (Stevens et al. 1999).

Tennant and coworkers (1994a) published several studies demonstrating that atrazine, at acute MTD oral doses, failed to stimulate uterine weight of OVX rats. Incorporation of [$^3$H]thymidine into uterine DNA, a more specific index of estrogen-mediated promotion of tissue growth, was similarly not stimulated by atrazine, nor was expression of the uterine progesterone receptor, known to be highly specific for, and sensitive to, estrogen. Studies by Connor et al. (1996) confirmed that atrazine did not induce uterine weight or progesterone receptor responses; they added a test of uterine peroxidase expression, which atrazine also failed to stimulate. Similarly, McMullin et al. (2004) reported that the uterine progesterone receptor mRNA was not stimulated by atrazine in OVX rats. Another established estrogen-mediated response in rats, the release of prolactin from pituitary tissue transplanted under the animal's kidney capsule, was examined by Fujimoto and Honda (2003); results showed no evidence that atrazine stimulated prolactin release.

Detailed histological examination was conducted on vaginal, uterine, and mammary tissues from intact SD female rats administered atrazine; no treatment-related estrogenic responses were observed (Eldridge et al. 1998). Eldridge et al. (1999b) provided a detailed analysis of vaginal cytology and estrous cycling patterns in intact SD rats administered atrazine. Within the first few weeks of dosing, vaginal cytology appeared less cornified (i.e., diminished estrogenic response), suggesting, once again, that atrazine was not acting as an estrogen agonist. Ashby et al. (2002) administered atrazine to prepubertal female rats and did not observe estrogen-dependent responses, i.e., stimulation of uterine weight. Tanaka et al. (2004) administered dimethylbenzanthracene (DMBA) with and without atrazine to stimulate mammary tumor growth in intact young female rats. Administration of DMBA initiated and/or promoted tumor growth. The incidence of ovarian and mammary tumors was not increased in the DMBA + atrazine treatment groups. In fact, a non-dose-responsive decrease of ovarian tumors was observed in all atrazine-treated groups. This study bears replication because no historical control data were provided, which prevents knowing, with certainty, if the control incidence of ovarian tumors in DMBA-treated control rats was elevated above the normal range.

A number of studies have also been conducted with atrazine in nonmammalian species. Crain et al. (1999) reported a lack of feminization of alligator eggs, or stimulation of the immature reproductive tract after atrazine exposure; both are very sensitive markers for estrogenic effects in reptiles. Similar results were recently reported for crocodile egg exposure by Beldomenico et al. (2007). De Solla et al. (2006) found that testicular development of snapping turtles was unaffected by incubation of eggs in soil containing atrazine. Although Hayes et al. (2002) reported that atrazine demasculinizes or feminizes developing male *Xenopus laevis* tadpoles (Hayes 2005; Hayes et al. 2006), Carr et al. (2003) did not find any effect of atrazine on the sex ratio of developing *Xenopus*. The result reported by Carr et al. (2003) was confirmed in a large study (Kloas et al. 2007) conducted concurrently in two laboratories. In this study, *Xenopus laevis* was exposed to atrazine at concentrations of 0.01, 0.1, 1.0, 25, or 100 ppb from day 8 postfertilization until the completion of metamorphosis; estradiol, administered under similar conditions, at a concentration of 0.2 ppb resulted in a significant increase in larvae with female or mixed sex gonads, compared to untreated controls. Bringof et al. (2004) reported that atrazine had no effect on a number of reproductive tract structures in minnows.

Production of fish vitellogenin, a well-established estrogen-mediated response, has been examined after atrazine exposure in goldfish (Spano et al. 2004) and carp (Sanderson et al. 2001); no effect of atrazine treatment was noted in either study. Atrazine also failed to stimulate vitellogenin mRNA production in incubations of zebrafish embryos (Muncke et al. 2007). Wilhelms et al. (2006) did not find any evidence that atrazine affected uterine weight or pituitary LH release in quail. Earlier, these authors had reported the absence of estrogen-like effects in the maturing reproductive tracts of male quail administered up to 1000 ppm atrazine (Wilhelms et al. 2005). Finally, a study by LeGoff and coworkers (2006) showed that atrazine did not induce the expression of yolk protein genes in *Drosophila* incubated on atrazine-containing medium, whereas estrogen was capable of inducing these genes.

In some studies, high doses of atrazine antagonized estrogen-mediated responses. For example, Tennant et al. (1994a) observed diminished uterine weight, diminished progesterone receptor expression, and diminished thymidine incorporation into uterine DNA in estrogen-treated OVX rats administered atrazine. Connor et al. (1996) reported a similar antagonism of the estrogen-mediated response in rat uterus. The concentration necessary to elicit the antagonistic responses were typically very high, at least $10^6$ times greater than the picomolar levels of estradiol needed to activate the estrogen receptor.

In contrast, a more recent study by McMullin et al. (2004) showed that atrazine did not block the estrogen-stimulated expression of mRNA that codes for the progesterone receptor. Sanderson et al. (2001) reported that carp exposed to atrazine did not display any effect on estrogen-stimulated vitellogenin production. In addition, the study by Wilhelms et al. (2005) on immature Japanese quail administered atrazine in the diet, and the study by LeGoff et al. (2006) on atrazine-exposed fruit flies, found no evidence of an effect of atrazine on naturally occurring estrogen-mediated responses.

Another well-studied estrogen-specific *in vivo* response in rodents is the ability of estrogen-primed OVX female rats to produce a daily surge of pituitary LH.

Administration of atrazine, for as short as 3 d, or as long as 6 mon, has been found to inhibit these estrogen-primed LH surges (Simpkins et al. 1998; Eldridge et al. 1999a; Cooper et al. 2000; McMullin et al. 2004). The successful generation of the light-entrained, estrogen-primed LH surge in rodents is dependent upon a cascade of several nonestrogen neuroendocrine components, any of which may be blocked by atrazine. Although possible, it is unlikely that atrazine antagonism of the estrogen receptor plays a role in the suppression of LH release.

In summary, from these *in vivo* studies, which employed a wide variety of well-recognized, standard, and specific biological responses to estrogen, it can be concluded that atrazine does not elicit estrogen-like responses, even at dose levels up to a million-fold greater than the minimally effective estrogen dose. These results support the conclusion that atrazine is not an estrogen receptor agonist.

In some of the previously described models, however, high doses of atrazine appeared to inhibit or reduce the response to estrogen. This "inhibition" typically occurs at atrazine doses near to, or greater than, the MTD, and at levels several orders of magnitude greater than the amount of estrogen required to initiate the response. Therefore, one can conclude, from the foregoing review, that atrazine antagonism of estrogen-mediated responses *in vivo* is either nonexistent or extremely weak, and is unlikely to be relevant to man under conditions of potential human exposure (US EPA 2006).

## 3.2 In Vitro Expression Systems

Table 2 presents an overview, taken from the literature, of responses to atrazine exposure during testing of estrogen expression *in vitro*.

A number of investigators have evaluated the interaction of atrazine with estrogen expression by using constructs containing an estrogen receptor (ER) and a reporter composed of an easy-to-measure cellular response, or with a reporter gene tethered to a naturally occurring estrogen-activated receptor-dependent genomic site. These *in vitro* models are relatively easy to run, highly precise and specific, and they permit assessment of both agonist and antagonist potential within the same system.

Atrazine has consistently failed to activate estrogen-dependent reporters *in vitro* in estrogen-dependent expression systems. Atrazine failed to stimulate estrogen-dependent MCF-7 cell proliferation (Soto et al. 1995; Connor et al. 1996; Fukamachi et al. 2004); and atrazine did not enhance an estrogen-induced increased aggregation of the progesterone receptor-progesterone response element (PR-PRE) derived from nuclear DNA extracts of MCF-7 cells (Connor et al. 1996). Estrogen-dependent proliferation of transfected yeast cells (Connor et al. 1996) and MtT/E-2 cells (Fujimoto and Honda 2003) did not occur upon incubation with atrazine. Similarly, production of specific estrogen-mediated products did not respond to atrazine in reporter constructs of MCF-7 cells (Connor et al. 1996; Balaguer et al. 1996), in HeLa cells transfected with the alpha- or beta-subunit forms of ER (Balaguer et al. 1996),

**Table 2** Responses to atrazine exposure using *in vitro* tests of estrogen expression

| Cell type | Test | Stimulated[a] | Inhibited[a,b] | Reference |
|---|---|---|---|---|
| MCF-7 | ER-mediated proliferation | No | No | Connor et al. 1996 |
| MCF-7 | ER-mediated proliferation | No | — | Soto et al. 1995 |
| MCF-7 | ER-mediated proliferation | No | — | Fukamachi et al. 2004 |
| MCF-7 | Nuclear DNA—PgR complex | No | No | Connor et al. 1996 |
| MCF-7 | ER-mediated genetic expression | No | No | Connor et al. 1996 |
| MCF-7 | ER-mediated genetic expression | No | No | Balaguer et al. 1996 |
| HeLa | ER-mediated genetic expression | No | — | Balaguer et al. 1996 |
| HeLa | ER-α- and ER-β-mediated expression | No | — | Balaguer et al. 1996 |
| CHO cells | ER-α- and ER-β-mediated expression | No | — | Kojima et al. 2004 |
| T47D.Luc | ER-mediated genetic expression | No | — | Legler et al. 2002 |
| MtT/E-2 | ER-mediated proliferation | No | — | Fujimoto and Honda 2003 |
| Yeast | ER-mediated proliferation | No | No | Connor et al. 1996 |
| Yeast | ER-mediated genetic expression | No | Weak | Tran et al. 1996 |
| Yeast | ER-mediated genetic expression | No | No | Graumann et al. 1999 |
| Yeast | ER-mediated genetic expression | No | — | O'Conner et al. 2000 |
| Clam gills | Metabolic stimulation | No | No | Cheney et al. 1997 |
| Fish. hepatic | Vitellogenin production | No | Weak | Sanderson et al. 2001 |

ER, estrogen receptor; —, assessment not done.
[a] No, indicates no response to atrazine was observed at doses that were at least $10^5$ in excess of the effective dose of estradiol.
[b] Weak, indicates a response to atrazine was observed only at doses that were at least $10^5$ in excess of the effective dose of estradiol.

in yeast cells (Tran et al. 1996; Graumann et al. 1999; O'Connor et al. 2000), or in T47D Luc cells (Legler et al. 2002).

Cheney and coworkers (1997) measured metabolic stimulation of clam gills incubated *in vitro* with atrazine, and Sanderson et al. (2001) incubated fish hepatic cells with atrazine; neither study produced evidence of an atrazine-induced "estrogen-like response." Other laboratories have tested for atrazine antagonism of estrogen stimulation in a transfected construct. Connor et al. (1996) reported that atrazine coincubated with estradiol failed to inhibit an ER-mediated proliferation of MCF-7 cells. Atrazine also failed to inhibit estrogen-mediated formation of a progesterone receptor complex with DNA, and showed a similar failure with ER-mediated genetic expression in MCF-7 cells (Connor et al. 1996). Balaguer et al. (1999) confirmed that atrazine did not inhibit ER-mediated expression in MCF-7 cells.

In studies of transfected yeast cells, Connor et al. (1996) found that atrazine did not block estrogen-mediated cell proliferation, and Graumann et al. (1999) reported that atrazine did not inhibit an ER expression system transfected into yeast. This latter result contrasted with earlier findings of Tran et al. (1996), who observed a weak inhibition by atrazine of an estrogen-mediated reporter in yeast cells. Atrazine was unable to block estrogen stimulation of clam gill metabolism (Cheney et al. 1997), but it weakly inhibited vitellogenin production by fish liver cells incubated *in vitro* (Sanderson et al. 2001).

To summarize the results from *in vitro* studies, a variety of tests using well-established, estrogen-dependent reporter systems have uniformly failed to demonstrate an ability of atrazine to imitate the action of estrogen. Moreover, a majority of tests of estrogen antagonism by atrazine also produced negative findings. We conclude from them foregoing that atrazine does not interact with the estrogen receptor in these incubation systems.

## 3.3   Estrogen Receptor Binding

Steroid hormones, such as estrogen, serve as an activating ligand for specific intracellular receptors (ER-$\alpha$ or ER-$\beta$). Upon activation, these hormone-bound proteins are attracted to specific chromatin sites. Once bound to DNA, the hormone–ligand complex serves to attract additional factors that initiate and maintain mRNA transcription. Although binding of a particular ligand to the estrogen receptor may not necessarily enhance or diminish an estrogen-mediated response, it is typically assumed that receptor interaction is mandatory for classic estrogen-mediated expression to occur. Many investigators have studied whether atrazine may interact directly with the estrogen receptor.

Table 3 presents a summary, from the literature, of studies in which there was atrazine competition against estrogen receptor binding in yeast or various animal tissues. Tennant et al. (1994b) reported that competitive coincubation of atrazine and radiolabeled estrogen with ER-containing rat uterine cytosol failed to produce significant displacement of estrogen binding by atrazine. However, when uterine cytosols were preincubated with atrazine, and a tritium tracer was later added to a chilled incubation, there was a significant reduction of [$^3$H]estradiol association. Scatchard analysis suggested a very weak competitive antagonism at concentrations that were several orders of magnitude greater than the subnanomolar disassociation constant ($K_d$) for estradiol binding to ER. Thomas and Dong (2006) recently reported that atrazine very weakly displaced estradiol binding to a seven-transmembrane receptor (GPR30) stably transfected into HEK-293 cells. These results indicate that, when atrazine is allowed to compete with estrogen for the estrogen receptor, no antagonism is found. However, when high concentrations of atrazine are preincubated with the estrogen receptor before the addition of estradiol, weak antagonism can be observed.

Tennant et al. (1994b) also showed that, when radiolabeled estradiol binding was assessed in uterine cytosols prepared from OVX rats orally dosed with atrazine

**Table 3** Atrazine competition against estrogen receptor (ER) binding in various tissues or organisms

| ER source | Incubation | Competition[a,b] | Reference |
|---|---|---|---|
| Rat uterus | In vitro | Weak | Tennant et al. 1994b |
| Rat uterus | In vitro | Weak | McMullin et al. 2004 |
| Rat uterus | In vitro | No | Danzo 1997 |
| Rat uterus | In vitro | No | O'Connor et al. 2000 |
| Rat uterus | Ex vivo | Weak | Tezak et al. 1992 |
| Rat uterus | Ex vivo | Weak | Tennant et al. 1994b |
| Rat hypothalamus | Ex vivo | No | McMullin et al. 2004 |
| Human ER-α and -β | In vitro | No | Roberge et al. 2004 |
| Human ER | In vitro | Weak | Hanoika et al. 1999 |
| Human ER-α | In vitro | Weak | Scippo et al. 2004 |
| Transfected yeast | In vitro | Weak | Tran et al. 1996 |
| Alligator oviduct | In vitro | Weak | Vonier et al. 1996 |
| ER GPR30 in HEK293 cells | In vitro | Weak | Thomas and Dong 2006 |

[a] Weak, indicates negative inhibition at concentrations of at least $10^5$ molar excess.
[b] No, indicates no inhibition at concentrations of at least $10^5$ molar excess.

before being killed, a similar reduction of radioligand association with its receptor was observed. This latter finding confirmed earlier work by Tezak et al. (1992). McMullin and coworkers (2004) observed a mild competition in rat hypothalamus between atrazine and estrogen for the estrogen receptor from orally administered atrazine. In contrast, other groups have not observed significant competition by atrazine for rat uterine ER binding *in vitro* (Danzo 1997; O'Connor et al. 2000; McMullin et al. 2004).

Hanioka et al. (1999) and Scippo et al. (2004), however, using a noncellular preparation containing recombinant human ER-α, observed a very weak competition for receptor binding by atrazine and estrogen, although Roberge et al. (2004) were unable to observe significant atrazine displacement of estradiol binding to incubated recombinant human ER-α or ER-β. Vonier and coworkers (1996) also reported that atrazine demonstrated limited competition against estradiol binding to cytosols prepared from alligator oviduct. Finally, Tran et al. (1996) observed very weak displacement of estradiol in yeast cells transfected with the estrogen receptor.

In conclusion, there is evidence that atrazine has a limited capacity to antagonize the binding of estrogen to its receptors. Antagonism of estrogen binding to ER is not expected to occur under equilibrium conditions, or at concentrations expected as a result of normal human exposure.

# 4 Summary

More than 40 publications have described results of atrazine responses in 17 estrogen-dependent systems and in more than a dozen different reporter and estrogen receptor-binding studies *in vitro*. Results from these studies have consistently failed to demonstrate that atrazine acts as an estrogen agonist. Moreover, a variety of indices of estrogen-dependent activity, in models that encompass cell incubations to whole animals, have failed to respond to atrazine. Researchers in more than a dozen laboratories have examined rats, rat tissues, human and prokaryotic cells, in addition to tissues from reptile, fish, amphibian, avian, molluscan, and insect sources, without eliciting estrogenic-like responses from atrazine.

In contrast, studies of atrazine ability to antagonize estrogen-mediated responses have yielded equivocal results. Results of several studies show inhibition of estrogen-like activities by atrazine, yet many other tests have yielded negative results. Generally, *in vivo* models have more consistently shown that atrazine inhibits estrogen-mediated responses, whereas in more specific *in vitro* systems, inhibition is seldom observed. The implication is that *in vivo* effects of atrazine may result from inhibition of factors that are indirectly connected to the genomic interaction of estrogen (e.g., at the receptor). Potential targets of atrazine may be downstream of the ligand–receptor binding event. Atrazine may also interact with other, less specific, factors that are necessary for the completion of the estrogen-mediated response.

Moreover, the apparent inhibition of cytosolic-ER binding by atrazine may, similarly, be relatively nonspecific. Observed inhibitory responses occur only at extreme doses or concentrations, i.e., several orders of magnitude greater than the level of estradiol presence in each test system. It is probable that the inhibitory effects result from very low affinity and/or low specificity interactions, which are unlikely to occur in nature.

We conclude that atrazine is not an estrogen receptor agonist, but it may be a weak antagonist, when present at a high concentration under conditions of disequilibrium with estrogen. These conditions are not expected to occur as a result of normal environmental exposure.

# References

APVMA (Australian Pesticides and Veterinary Medicines Authority) (2004) The reconsideration of approvals of the active constituent atrazine, registrations of products containing atrazine, and their associated labels. Second Draft Final Review Report Including Additional Assessments. October 2004. Australian Pesticides and Veterinary Medicines Authority, Canberra, Australia.

Ashby J, Tinwell H, Stevens J, Pastoor T, Breckenridge CB (2002) The effects of atrazine on the sexual maturation of female rats. Regul Toxicol Pharm 35:468–473.

Balaguer P, Joyeux A, Denison MS, Vincent R, Gillesby BE, Zacharewski T (1996) Assessing the estrogenic and dioxin-like activities of chemicals and complex mixtures using in vitro recombinant receptor-reporter gene assays. Can J Physiol Pharmacol 74:216–222.

Balaguer P, Francois F, Comunale F, Fenet H, Boussioux AM, Pons M, Nicolas JC, Casellas C (1999) Reporter cell lines to study the estrogenic effects of xenoestrogens. Sci Total Environ 233:47–56.

Beldomenico PM, Rey F, Prado WS, Villarreal JC, Munoz-de-Toro M, Luque EH (2007) *In ovum* exposure to pesticides increases the egg weight loss and decreases hatchlings weight of *Caiman latirostris* (Crocodylia: Alligatoridae). Ecotoxicol Environ Saf 68:246–251.

Bringolf RB, Belden JB, Summerfelt RC (2004) Effects of atrazine on fathead minnow in a short-term reproduction assay. Environ Toxicol Chem 23:1019–1025.

Calle EE, Frumkin H, Henley SJ, Savitz DA, Thun MJ (2002) Organochlorines and breast cancer risk. CA Cancer J Clin 52:301–309.

Carr JA, Gentiles A, Smith EE, Goleman WL, Urquidi LJ, Thuett K, Kendall RJ, Giesy JP, Gross TS, Solomon KR, Van Der Kraak G (2003) Response of larval *Xenopus laevis* to atrazine: assessment of growth, metamorphosis, and gonadal and laryngeal morphology. Environ Toxicol Chem 22:396–405.

Cheney MA, Fiorillo R, Criddle RS (1997) Herbicide and estrogen effects on the metabolic activity of *Elliptio complanata* measured by calorespirometry. Comp Biochem Physiol C Pharmacol Toxicol Endocrinol 118:159–164.

Conner K, Howell J, Chen I, Liu H, Berhane K, Sciarretta C, Safe S, Zacharewski T (1996) Failure of chloro-s-triazine-derived compounds to induce estrogen receptor-mediated responses *in vivo* and *in vitro*. Fundam Appl Toxicol 30:93–101.

Cooper RL, Stoker TE, Tyrey L, Goldman JM, McElroy WK (2000) Atrazine disrupts the hypothalamic control of pituitary-ovarian function. Toxicol Sci 53:297–307.

Crain DA, Spiteri ID, Guillette LJ Jr (1999) The functional and structural observations of the neonatal reproductive system of alligators exposed *in ovo* to atrazine, 2,4-D, or estradiol. Toxicol Ind Health 15:180–185.

Danzo B (1997) Environmental xenobiotics may disrupt normal endocrine function by interfering with the binding of physiological ligands to steroid receptors and binding proteins. Environ Health Perspect 105:294–301.

Davis DL, Bradlow HL, Wolff M, Woodruff T, Hoel DG, Anton-Culver H (1993) Medical hypothesis: xenoestrogens as preventable causes of breast cancer. Environ Health Perspect 101:372–377.

De Solla SR, Martin PA, Fernie KJ, Park BJ, Mayne G (2006) Effects of environmentally relevant concentrations of atrazine on gonadal development of snapping turtles (*Chelydra serpentine*). Environ Toxicol Chem 25:520–526.

Devine MD, Duke SO, Fedtke C (1993) Herbicidal inhibition of photosynthetic electron transport. In: Physiology of Herbicide Action. Prentice-Hall, Englewood Cliffs, NJ, pp 113–140.

Eldridge JC, Tennant MK, Wetzel LT, Breckenridge CB, Stevens JT (1994) Factors affecting mammary tumor incidence in chlorotriazine-treated female rats: hormonal properties, dosage and animal strain. Environ Health Perspect 102(suppl 11):29–36.

Eldridge JC, McConnell RF, Wetzel LT, Tisdel, MO (1998) Appearance of mammary tumors in atrazine-treated female rats: probable mode of action involving strain-related control of ovulation and estrous cycling. In: Ballantine LG, McFarland JE, Hackett DS (eds) Triazine Herbicides: Risk Assessment. Oxford University Press, Washington, DC, pp 414–423.

Eldridge JC, Wetzel LT, Stevens JT, Simpkins JW (1999a) The mammary tumor response in triazine-treated female rats: a threshold-mediated interaction with strain and species-specific reproductive senescence. Steroids 64:672–678.

Eldridge JC, Wetzel LT, Tyrey L (1999b) Estrous cycle patterns of Sprague–Dawley rats during acute and chronic atrazine administration. Reprod Toxicol 13:491–499. European Union (2000) Atrazine. Volume 3. Annex B. Addendum to the Report and Proposed Decision of the United Kingdom Made to the European Commission under Article 7(1) of Regulation 3600/92. Summary, Scientific Evaluation and Assessment. Council Directive 91/414/EEC Regulation 3600/92. February 2000.

Fujimoto N, Honda H (2003) Effects of environmental estrogenic compounds on growth of a transplanted estrogen responsive pituitary tumor cell line in rats. Food Chem Toxicol 41:1711–1717.

Fukamachi K, Han BS, Kim CK, Takasuka N, Matsuoka Y, Matsuda E, Yamasaki T, Tsuda H. (2004) Possible enhancing effects of atrazine and nonylphenol on 7,12-dimethylbenz[*a*]anthracene induced mammary tumor development in human c-Ha-ras proto-oncogene transgenic rats. Cancer Sci 95:404–410.

Good NE (1961) Inhibitors of the Hill reaction. Plant Physiol 36:788–803.

Graumann K, Breithofer A, Jungbauer A (1999) Monitoring of estrogen mimics by a recombinant yeast assay: synergy between natural and synthetic compounds? Sci Total Environ 225:69–79.

Hanioka N, Jinno H, Tanaka-Kagawa T, Nishimura T, Ando M (1999) In vitro metabolism of simazine, atrazine and propazine by hepatic cytochrome P450 enzymes of rat, mouse and guinea pig, and oestrogenic activity of chlorotriazines and their main metabolites. Xenobiotica 29:1213–1226.

Hauswirth JW, Wetzel LT (1998) Toxicity characteristics of the 2-chlorotriazines atrazine and simazine. In: Ballantine LG, McFarland JE, Hackett DS (eds) Triazine Herbicides: Risk Assessment. Oxford University Press, Washington, DC, pp 370–383.

Hayes TB (2005) Atrazine and pesticide mixtures: sum of the parts or some of the parts? In: SETAC Annual Meeting, Baltimore, MD, USA. SETAC, Pensacola, FL.

Hayes TB, Collins A, Lee M, Mendoza M, Noriega N, Stuart AA, Vonk A (2002) Hermaphoditic, demasculinized frogs after exposure to the herbicide atrazine at low ecologically relevant doses. Proc Natl Acad Sci U S A 99(8):5476–5480.

Hayes TB, Stuart AA, Mendoza M, Collins A, Noriega N, Vonk A, Johnston G, Liu R, Kpodzo D (2006) Characterization of atrazine-induced gonadal malformations in African clawed frogs (*Xenopus laevis*) and comparisons with effects of an androgen antagonist (cyproterone acetate) and exogenous estrogen (17-estradiol): support for the demasculinization/feminization hypothesis. Environ Health Perspect 114(1):134–141.

IARC (1999) Atrazine. Some Chemicals That Cause Tumours of the Kidney or Urinary Bladder in Rodents and Some Other Substances. World Health Organization International Agency for Research on Cancer. IARC Monographs on the Evaluation of Carcinogenic Risks to Humans. Views and Expert Opinions of an IARC Working Group on the Evaluation of Carcinogenic Risks to Humans, Lyon, 13–20 October 1998, 33:59–113.

Kloas W, Lutz I, Springer T, Krueger H, Wolf J, Holden L, Hosmer A (2008) Atrazine does not induce gonadal feminization in *Xenopus laevis*. Submitted to Proc Natl Acad Sci U S A.

Kojima H, Katsura E, Takeuchi S, Niiyama K, Kobayashi K (2004) Screening for estrogen and androgen receptor activities in 200 pesticides by *in vitro* reporter gene assays using Chinese hamster ovary cells. Environ Health Perspect 112:524–531.

Legler J, Dennenkamp M, Vethaak AD, Brouwer A, Koeman JH, van der Burg B, Murk AJ (2002) Detection of estrogenic activity in sediment-associated compounds using *in vitro* reporter gene assays. Sci Total Environ 293:69–83.

LeGoff G, Hilliou F, Siegfried BD, Boundy S, Wajnberg E, Sofer L, Audant P, French-Constant RH, Feyereisen R (2006) Xenobiotic response in *Drosophila melanogaster*: sex dependence of P450 and GST gene induction. Insect Biochem Mol Biol 36:674–682.

McMullin TS, Andersen ME, Nagahara A, Lund TD, Pak T, Handa RJ, Hanneman WH (2004) Evidence that atrazine and diaminochlorotriazine inhibit the estrogen/progesterone induced surge of luteinizing hormone in female Sprague–Dawley rats without changing estrogen receptor action. Toxicol Sci 79:278–286.

Mitra AK, Faruque FS, Avis AL (2004) Breast cancer and environmental risk: where is the link? J Environ Health 66:24–32.

Muir K, Rattanamongkolgul S, Smallman-Raynor M, Thomas M, Downer S, Jenkinson C (2004) Breast cancer incidence and its possible spatial association with pesticide application in two counties of England. Public Health 118:513–520.

Mukherjee S, Koner BC, Ray S, Ray A (2006) Environmental contaminants in pathogenesis of breast cancer. Indian J Exp Biol 44:597–617.

Muncke J, Junghans M, Eggen RIL (2007) Testing estrogenicity of known and novel (xeno-) estrogens in the MolDarT using developing zebrafish (*Danio rerio*). Environ Toxicol 22:185–193.

O'Connor JC, Plowchalk DR, Van Pelt CS, Davis LG, Cook JC (2000) Role of prolactin in chloro-*s*-triazine rat mammary tumorigenesis. Drug Chem Toxicol 23:575–601.

Qiao X, Ma L, Hummel HE (1996) Persistence of atrazine and occurrence of its primary metabolites in three soils. J Agric Food Chem 44:2846–2848.

Roberge M, Hakk H, Larsen G (2004) Atrazine is a competitive inhibitor of phosphodiesterase but does not affect the estrogen receptor. Toxicol Lett 154:61–68.

Safe S (2004) Endocrine disruptors and human health: is there a problem? Toxicology 205:3–10.

Sanderson JT, Letcher RJ, Heneweer M, Giesy JP, van den Berg M (2001) Effects of chloro-s-triazine herbicides and metabolites on aromatase activity in various human cell lines and on vitellogenin production in male carp hepatocytes. Environ Health Perspect 109:1027–1031.

Schottler SP, Eisenreich SJ, Hines NA, Warren G (1998) Temporal and spatial trends of atrazine, desethylatrazine, and desisopropylatrazine in the Great Lakes. In: Ballantine LG, McFarland JE, Hackett DS (eds) Triazine Herbicides: Risk Assessment. Oxford University Press, Washington, DC, pp 208–226.

Scippo ML, Argiris C, Van De Weerdt C, Muller M, Willemsen P, Martial J, Maghuin-Rogister G (2004) Recombinant human estrogen, androgen and progesterone receptors for detection of potential endocrine disruptors. Anal Bioanal Chem 378:664–669.

Simpkins JW, Eldridge JC, Wetzel LT (1998) Role of strain-specific reproductive patterns in appearance of mammary tumors in atrazine treated rats. In: Ballantine LG, McFarland JE, Hackett DS (eds) Triazine Herbicides: Risk Assessment. Oxford University Press, Washington, DC, pp 399–413.

Sonnenschein C, Soto AM (1998) An updated review of environmental estrogen and androgen mimics and antagonists. J Steroid Biochem Mol Biol 65:143–150.

Soto AM, Sonnenschein C, Chung KL, Fernandez MF, Olea N, Serrano FO (1995) The E-screen assay as a tool to identify estrogens: an update on estrogenic environmental pollutants. Environ Health Perspect 104(suppl 7):113–122.

Spano L, Tyler CR, van Aerle R, Devos P, Mandiki SN, Silvestre F, Thome JP, Mestemont P (2004) Effects of atrazine on sex steroid dynamics, plasma vitellogenin concentration and gonadal development in adult goldfish (*Carassius auratus*). Aquat Toxicol 66:369–379.

Steingraber S (1997) Mechanisms, proof, and unmet needs: the perspective of a cancer activist. Environ Health Perspect 105(suppl 3):685–687.

Stevens JT, Breckenridge CB, Wetzel L (1999) A risk characterization for atrazine. J Toxicol Environ Health 56:69–109.

Tanaka T, Kohno H, Suzuki R, Sugie S (2004) Lack of modifying effects of an estrogenic compound atrazine on 7,12-dimethylbenz(*a*)anthracene-induced ovarian carcinogenesis in rats. Cancer Lett 210:129–137.

Tennant MK, Hill DS, Eldridge JC, Wetzel LT, Breckenridge CB, Stevens JT (1994a) Possible antiestrogenic properties of chloro-*s*-triazines in rat uterus. J Toxicol Environ Health 43:183–196.

Tennant MK, Hill DS, Eldridge JC, Wetzel LT, Breckenridge CB, Stevens JT (1994b) Chloro-*s*-triazine antagonism of estrogen action: limited interaction with estrogen receptor binding. J Toxicol Environ Health 43:197–211.

Tezak Z, Simi B, Kniewald J (1992) Effect of pesticides on oestradiol-receptor complex formation in rat uterus cytosol. Food Chem Toxicol 30:879–885.

Thomas P, Dong J (2006) Binding and activation of the seven-transmembrane estrogen receptor GPR30 by environmental estrogens: a potential novel mechanism of endocrine disruption. J Steroid Biochem Mol Biol 102:175–179.

Tran DQ, Kow KY, McLachlan JA, Arnold SF (1996) The inhibition of estrogen receptor-mediated responses by chloro-*s*-triazine-derived compounds is dependent on estradiol concentration in yeast. Biochem Biophys Res Commun 227:140–146.

United Kingdom Pesticide Directorate (2000) Atrazine: Report and Proposed Decision of the United Kingdom Made to the European Commission Under Article 7(1) of Regulation 3600/92 Council Directive 91/4 14/EEC Regulation 3600/92, UK. Rapporteur Monograph, October 1996.

US EPA (1996) Drinking water regulation and health advisories. EPA 822-B-96-002. Office of Water, U.S. Environmental Protection Agency, Washington, DC.

US EPA (2002) Atrazine Toxicology Chapter of the Re-registration Eligibility Decision, Second Revision, April 11, 2002.

US EPA (2006) Triazine Cumulative Risk Assessment. HED Human Health Risk Assessment in Support of the Reregistration Eligibility Decisions for Atrazine, Simazine and Propazine. PC Codes: 080808, 080803, 080807. DP 317976. Health Effects Division (7509C), Benefits and Economic Analysis Division (7503C), and Environmental Fate and Effects Division (7507C). March 28, 2006. Office of Prevention, Pesticides and Toxic Substances, United States Environmental Protection Agency, Washington, DC.

Vencill WK (2002) Herbicide Handbook, 8th Ed. Weed Science Society of America, Lawrence, KS, pp 493–494.

Vonier PM, Crain DA, McLachlan JA, Guillette LJ Jr, Arnold SF (1996) Interaction of environmental chemicals with the estrogen and progesterone receptors from the oviduct of the American alligator. Environ Health Perspect 104:1318–1322.

Wetzel LT, Luempert LG III, Breckenridge CB, Tisdel MO, Stevens JT, Thakur AK, Extrom PJ, Eldridge JC (1994) Chronic effects of atrazine on estrus and mammary tumor formation in female Sprague-Dawley and Fischer-344 rats. J Toxicol Environ Health 43:169–182.

Wilhelms KW, Cutler SA, Proudman JA, Anderson LL, Scanes CG (2005) Atrazine and the hypothalamo-pituitary-gonadal axis in sexually maturing precocial birds: studies in the male Japanese quail. Toxicol Sci 86:152–160.

Wilhelms KW, Cutler SA, Proudman JA, Carsia RV, Anderson LL, Scanes CG (2006) Lack of effects of atrazine on estrogen-responsive organs and circulating hormone concentrations in sexually immature female Japanese quail (*Coturnix coturnix japonica*). Chemosphere 3:64

Winkelmann DA, Kliane SJ (1990) Degradation and bound residue formation of urof atrazine metabolites, deethylatrazine, dealkylatrazine and hydroxyatrazine, in a western Tennessee soil. Environ Toxicol Chem 10(3):347–354.

# Index

## A

Aquatic animal toxicity, carbaryl (table), 108
Aquatic ecosystem contamination, OCPs (organochlorine pesticides), 18
Aquatic plant residues, pesticides, 22
Aquatic plant residues, xenobiotics (table), 24
Aromatic plants, pesticide residues (table), 33
Arsenic air emission standards, Chile (table), 141
Arsenic air pollution, reduction costs, 139
Arsenic contaminated media, Chile (table), 128
Arsenic exposure in Chile, geographic distribution (table), 129
Arsenic exposure in Chile, pathway (table), 131
Arsenic exposure pathway, Chilean cities (table), 130
Arsenic exposure, Chilean drinking water (table), 131
Arsenic health effects, Chile, 131
Arsenic in air, Chile (table), 126
Arsenic in Chile, human exposure, 129
Arsenic in Chile, polluting smelters (illus.), 124
Arsenic in Chile, reducing airborne emissions, 137
Arsenic in soils and vegetables, Chile, 127
Arsenic in water, Chile (table), 127
Arsenic levels in soil, Chilean cities (illus.), 128
Arsenic levels in vegetables and fruits (illus.), 128
Arsenic pollution in Chile, environmental media levels, 125
Arsenic pollution, Chilean environmental standards, 140
Arsenic reduction, process efficiencies (table), 138
Arsenic removal costs, Chilean water, 135
Arsenic removal from drinking water, cost (illus.), 136
Arsenic water mitigation technologies, Chile, 132
Arsenic, hazardous pollutant in Chile, 123 ff.
Arsenic, limits in Chilean water (table), 142
Arsenic-removal facilities, Chile (table), 134
Arsenic-removal, Chilean surface- and ground-water (illus.), 133
Atrazine in vitro studies, estrogen-mediated expression, 152
Atrazine in vitro tests, estrogen expression (table), 153
Atrazine in vivo studies, estrogen-mediated expression (table), 149
Atrazine interaction, estrogen expression systems, 147 ff.
Atrazine, characteristics, 147
Atrazine, competition against estrogen receptor (table), 155
Atrazine, effects on rodent hypothalamic-pituitary-gonadal (HPG) axis, 148
Atrazine, environmental estrogen, 148
Atrazine, estrogen receptor binding, 154

## B

Biodegradation, defluorination, 64
Biodegradation, organohalogens, 56
Biodegradation, perfluorinated compounds, 53 ff., 60
Biodegradation, perfluorinated surfactants, 61
Biodegradation, polyfluorinated compounds (PFCs), 60
Bird residues, pesticides, 22
Bird residues, xenobiotics (table), 24

## C

Carbamate pesticide residues, environment (table), 23
Carbaryl metabolism, cytochrome P450 dependence (illus.), 106

Carbaryl metabolism, higher organisms, 106
Carbaryl toxicity, animals, 108
Carbaryl toxicity, humans, 112
Carbaryl, abiotic degradation, 101
Carbaryl, acute animal toxicity (table), 109
Carbaryl, aquatic animal toxicity (table), 108
Carbaryl, biotic degradation, 104
Carbaryl, chemistry, 96
Carbaryl, chemodynamics in air, 97
Carbaryl, chemodynamics in soil, 98
Carbaryl, chemodynamics in water, 98
Carbaryl, chronic animal toxicity, 110
Carbaryl, degradation pathways (illus.), 102
Carbaryl, environmental fate and toxicity, 95 ff.
Carbaryl, mammalian toxicokinetics and metabolism, 113
Carbaryl, microbial degradation pathway (illus.), 105
Carbaryl, photolytic pathway (illus.), 103
Carbaryl, physiochemical properties (table), 97
Carbaryl, reproductive toxicity, 111
Carbaryl, soil adsorption (table), 100, 101
Carbaryl, subacute animal toxicity, 109
Carbaryl, toxicity to insects and aquatic organisms, 107
Carbaryl, water residues (table), 99
Cereal grain contamination, pesticides (table), 27
Chemical food contamination, Egypt, 24
Chemodynamics in air, carbaryl, 97
Chemodynamics in soil, carbaryl, 98
Chemodynamics in water, carbaryl, 98
Children urine residues, OCP residues (table), 35
Chile, annual cost to remove arsenic (illus.), 136
Chile, arsenic pollution, 123 ff.
Chile, human exposure to arsenic, 129
Chilean air pollution, arsenic reduction costs, 139
Chilean air standards, arsenic (table), 141
Chilean airborne emission reduction, arsenic, 137
Chilean arsenic exposure, pathway contribution (table), 131
Chilean arsenic pollution, environmental media levels, 125
Chilean arsenic pollution, setting environmental standards, 140
Chilean arsenic pollution, smelter locations (illus.), 124
Chilean arsenic-removal process, schematic (illus.), 133
Chilean cities, arsenic exposure (table), 130
Chilean cities, arsenic in soil (illus.), 128
Chilean cities, atmospheric arsenic (table), 126
Chilean health effects, arsenic exposure, 131
Chilean population, arsenic exposure (table), 129
Chilean regulation of water, arsenic (table), 142
Chilean rivers, arsenic levels (table), 127
Chilean vegetables and fruits, arsenic levels (illus.), 128
Chilean water pollution, arsenic mitigation technologies, 132
Chilean water treatment facilities, arsenic removal (table), 134
Chilean water, arsenic removal costs, 135
Chilean water, characteristics (table), 133
Chlorinated pesticides, in dairy products (table), 26
Citrus residues, pesticides (table), 28
Cotton insecticide use, Egypt (table), 3
Cotton leafworm resistance, insecticides (table), 11
Cotton pesticide consumption, Egypt (table), 4
Cotton pests, Egypt, 3
Cytochrome P450 dependence, carbaryl metabolism (illus.), 106

**D**

Dairy product contamination, pesticides (table), 26
DDE blood residues, Egyptian women (table), 34
DDT milk contamination, Egypt, 25
Decrees, Egyptian pesticide control, 5
Defluorination, biodegradation, 64
Degradation in activated sludge, PFCs (perfluorinated compounds), 63
Degradation in sediment, PFCs, 63
Degradation of carbaryl, abiotic, 101
Degradation of carbaryl, biotic, 104
Degradation pathway, 2(N-ethyl perfluoroocta nesulfonarnido)ethanol (N-EtFOSE), activated sludge (illus.), 63
Degradation pathway, 8-2 fluorotelomer alcohol (illus.), 62
Degradation pathways, carbaryl (illus.), 102
Dehalogenation, energy implications, 58
Dehalogenation, Gibbs free energy (table), 59
Dehalogenation, organohalogens, 58
Dietary exposure to pesticides, health risks (table), 40
Dihaloelimination, Gibbs free energy changes (table), 67
Drinking water exposure in Chile, arsenic (table), 131

Index 163

Drinking water in Chile, arsenic remediation options (table), 135
Drinking water residues, pesticides (table), 15

**E**

Effects in plants, lead, 76
Egypt, agricultural pests, 3
Egypt, environmental pesticide impact, 1 ff.
Egypt, farm animal poisoning, 10
Egypt, hazardous pesticides, 4, 5
Egypt, insecticide consumption (table), 4
Egypt, insecticide use, 3
Egypt, methyl parathion spill, 11
Egypt, non-drug poisoning (illus.), 10
Egypt, organochlorine (OC) residues (table), 12, 14
Egypt, PAH pollution, 6
Egypt, pesticide distribution by class, 4
Egypt, pesticide handling behavior (table), 43, 45
Egypt, pesticide handling risks, 40
Egypt, pesticide poisoning, (table), 7
Egypt, pesticide toxicity in humans, 6
Egypt, poisoning statistics, 8
Egypt, pollution, 2
Egypt's aquatic ecosystem, OCP accumulation (table), 20
Egypt's aquatic ecosystem, OCP residues (table), 20
Egyptian aquatic ecosystems, PCBs (table), 21, 22
Egyptian Decrees, pesticides, 5
Egyptian dietary risk, pesticides, 38
Egyptian fish contamination, organochlorines (table), 17
Egyptian food contamination, non-pesticides, 22
Egyptian food, pesticide contamination, 25
Egyptian homes, pesticide exposure (table), 7
Egyptian pesticides, environmental impact, 5
Egyptian potatoes and citrus, pesticide residues (table), 28
Egyptian residues, pesticides in plants and birds, 22
Egyptian soil and vegetables, insecticide residues (table), 31
Environmental contamination, lead, 74
Environmental distribution, PFCs, 56
Environmental estrogen, atrazine, 148
Environmental fate, carbaryl, 95 ff.
Environmental impact, Egyptian pesticides, 5
Environmental impact, pesticides, 1 ff.
Environmental pollution sources, lead (illus.), 75
Environmental residues, pesticides (table), 23
Environmental standards in Chile, controlling arsenic pollution, 140
Estrogen expression systems, atrazine, 147 ff.
Estrogen expression systems, in vitro atrazine tests (table), 153
Estrogen receptor binding, atrazine, 154
Estrogen receptor competition, atrazine (table), 155
Estrogen-mediated expression, in vitro atrazine studies, 152
Estrogen-mediated expression, in vivo atrazine studies (table), 149
Expression systems, atrazine and estrogens, 147 ff.

**F**

Farm animal poisoning, Egypt, 10
Farm worker exposure, pesticides, 8
Fish consumption, pesticide risks (table), 37
Fish contamination, OC chemicals, 17
Fish residues in Mediterranean Sea, OCPs (table), 19
Fish residues in the Red Sea, OCs (table), 19
Fish residues, OCPs (illus.), 18
Fluorinated ethanes, Gibbs free energy changes (illus.), 67
Fluorinated propanes, thermodynamic properties (table), 65
Foliar deposition in plants, lead, 76
Food arsenic residues, Chile (illus.), 128
Fruit pesticide residues, Egypt (table), 29, 30
Fruit residues, pesticides (table), 39
Fruits, pesticide intake risks (table), 40
Fruits, pesticide residue intake (table), 38

**G**

Gibbs free energy changes, fluorinated ethanes (illus.), 67
Gibbs free energy changes, hydrogenolysis and dihaloelimination (table), 67
Gibbs free energy, dehalogenation (table), 59
Gibbs free energy, reductive fluorination (illus.), 66
Groundwater, arsenic-removal process (illus.), 133

**H**

Halogenated compounds, microbial cometabolism, 60
Hazardous pesticides, Egypt, 4

Hazardous pollutant, arsenic in Chile, 123 ff.
Health effects, arsenic in Chile, 131
Health risks, dietary pesticides (table), 40
Higher plants, lead stress effects, 73 ff.
HPG (hypothalamic-pituitary-gonadal) axis, atrazine effects, 148
Human blood residues, DDE and PCBs (table), 34
Human fluids, pesticide residues, 33
Human poisoning, Egypt (illus.), 9
Human poisoning, OP insecticides, 10
Human poisoning, pesticides (table), 7
Human poisonings, Egypt (table), 9
Human risk, OC insecticide fish residues (table), 37
Human toxicity, carbaryl, 112
Human toxicity, pesticides, 6
Hydrogenolysis, Gibbs free energy changes (table), 67

## I

In vitro estrogen expression, atrazine tests (table), 153
In vitro expressions systems, atrazine studies, 152
In vivo expressions systems, atrazine studies, 149
Infants, OCP residues in breast milk (table), 36
Infants, pesticides in breast milk, 35
Insect resistance, Egypt, 11
Insecticide residues, vegetables and soil (table), 31
Insecticide resistance, cotton leafworm (table), 11
Insecticide toxicity in humans, Egypt, 6
Insecticide use, Egyptian cotton (table), 3
Insecticides, Egypt, 3

## L

Lead absorption and transport, plants, 78
Lead accumulation, environment, 74
Lead accumulation, plants, 78
Lead behavior and effects, plants, 76
Lead contamination, environment, 74
Lead effect on nitrogen assimilation, plants, 85
Lead effects, physiobiochemical plant activities, 73 ff.
Lead effects, plant enzymes (table), 83
Lead effects, plant metabolic processes (table), 85
Lead effects, plant photosynthesis, 80
Lead effects, plant pigment control, 86
Lead effects, plant processes (illus.), 82
Lead localization, plants, 77
Lead pollution, sources (illus.), 75
Lead stress effects, higher plants, 73 ff.
Lead toxicity, organisms, 74
Lead toxicity, plants (illus.), 79
Lead uptake, plants, 76
Lead, effects on plant growth, 80
Lead, effects on seeds and seedlings, 80
Lead, plant respiration effects, 84
Lead, plant toxicity, 77
Leptophos poisoning, farm animals, 10

## M

Meat contamination, pesticides (table), 27
Medicinal plants, pesticide residues (table), 32, 33
Metabolic process effects, lead (table), 85
Metabolism in higher organisms, carbaryl, 106
Metabolism of carbaryl, mammals, 113
Methyl parathion spill, Port-Said, 11
Microbial cometabolism, halogenated compounds, 60
Microbial defluorination, biodegradation, 64
Microbial degradation pathway, carbaryl (illus.), 105

## N

Nile river contamination, fish residues (illus.), 18
Nile River pollution, pesticide residues (illus., table), 13, 14
Nitrogen assimilation of plants, lead effects, 85

## O

OC (organochlorine) water residues, Egypt (table), 12
OC chemicals, fish contamination, 17
OC insecticide residues, vegetables and soil (table), 31
OC insecticide risks, fish consumption (table), 37
OC residues in fish, Red Sea (table), 19
OC residues, organically farmed soils (table), 31
Occupational pesticide exposure, Egypt, 6, 7
OCP (organochlorine pesticides), sediment contamination, 15
OCP fish residues, Mediterranean Sea (table), 19
OCP residue accumulation, Egypt's aquatic ecosystems (table), 20

Index 165

OCP residues in infants, breast milk (table), 36
OCP residues, aquatic ecosystems, 18
OCP residues, Egypt's aquatic ecosystems (table), 20
OCP residues, Gisa school children (table), 35
OCP residues, Nile River fish (illus.), 18
OCs, ecosystem distribution, 20
OP (organophosphate) insecticide poisoning, farm animals, 10
OP insecticide residues, environment (table), 23
Organochlorine contamination, fish (table), 17
Organochlorine pesticides, Nile River residues (illus., table), 13, 14
Organochlorine residues, sediments (table, illus.), 16
Organochlorine water residues, Egypt (table), 12
Organohalogens, biodegradation, 56
Organohalogens, dehalogenation, 58
Organophosphate (OP) insecticide poisoning, farm animals, 10

## P

PAHs (polyaromatic hydrocarbons), pollution, 6
PCB (polychlorinated biphenyls) blood residues, Egyptian women (table), 34
PCBs, Egyptian aquatic ecosystem (table), 21, 22
PCBs, sediment contamination, 15
Perfluorinated compounds (PFCs), environmental distribution, 56
Perfluorinated surfactants, biodegradation, 61
Pesticide active ingredients, Egypt (table), 4
Pesticide consumption, Egypt (table), 4
Pesticide contamination, dairy and food products (table), 26
Pesticide contamination, Egyptian meats and liver (table), 27
Pesticide contamination, sediments, 15
Pesticide distribution by class, Egypt, 4
Pesticide exposure, Egyptian homes (table), 7
Pesticide exposure, farm workers, 8
Pesticide food contamination, Egypt, 25
Pesticide handling behavior, Egypt (table), 43, 45
Pesticide handling risks, Egypt, 40
Pesticide intake, fruits and vegetables, 37
Pesticide market, Egypt, 2
Pesticide occupational exposure, effects, 7, 8
Pesticide poisoning, Egypt (table), 7
Pesticide residues, aromatic plants (table), 33
Pesticide residues, cereal grains (table), 27
Pesticide residues, human fluids, 33

Pesticide residues, medicinal and aromatic plants, 31
Pesticide residues, medicinal and aromatic plants, 31
Pesticide residues, medicinal plants (table), 32, 33
Pesticide residues, vegetables and fruits (table), 39
Pesticide residues, water, 13
Pesticide risk, Egyptian diet, 38
Pesticide risks, fish consumption (table), 37
Pesticide toxicity in humans, Egypt, 6
Pesticide uptake, breast-fed infants, 35
Pesticide use, Egypt, 3
Pesticides and pollution, Egypt, 2
Pesticides in potatoes and citrus, Egypt (table), 28
Pesticides in vegetables and fruit, Egypt (table), 29, 30
Pesticides, aquatic plants and bird residues, 22
Pesticides, Cairo drinking water (table), 15
Pesticides, environmental impact, 1 ff.
PFCs (perfluorinated compounds), environmental distribution, 56
PFCs, biodegradation, 53 ff., 60
PFCs, examples (table), 54
PFCs, production process (illus.), 55
PFCs, properties, 54
Phosvel poisoning, farm animals, 10
Photolytic degradation pathway, carbaryl (illus.), 103
Photosynthesis, lead effects, 80
Physiochemical properties, carbaryl (table), 97
Pigment control in plants, lead effects, 86
Plant enzyme activity, lead effects (table), 83
Plant growth, lead effects, 80
Plant metabolic processes, lead effects (table), 85
Plant photosynthesis, lead effects, 80
Plant process effects, lead (illus.), 82
Plant respiration, lead effects, 84
Plants, lead absorption and transport, 78
Plants, lead accumulation, 78
Plants, lead deposition, 76
Plants, lead distribution, 77
Plants, lead effects on nitrogen assimilation, 85
Plants, lead exposure effects, 76
Plants, lead toxicity (illus.), 79
Plants, lead uptake, 76
Poisoning in Egypt, humans (table), 9
Poisoning in Egypt, non-drug (illus.), 10
Poisoning incidence, Egypt (illus.), 9
Poisoning statistics, Egypt, 8
Poisoning, pesticides in Egypt (table), 7

Pollutant in Chile, arsenic, 123 ff.
Polyaromatic hydrocarbons (PAHs), pollution, 6
Polyfluorinated compounds, biodegradation, 60
POPs (priority organic pollutants) food contamination, Egypt, 22
Potato residues, pesticides (table), 28
Production process, perfluorinated compounds (illus.), 55

**R**

Reducing arsenic pollution, process efficiencies (table), 138
Reductive dechlorination, vitamin role, 60
Reductive fluorination, Gibbs free energy (illus.), 66
Reproductive toxicity, carbaryl, 111
Respiration in plants, lead effects, 84

**S**

Sediment contamination, PCBs, 15
Sediment OC residues, Egypt (table, illus.), 16
Sediment, PFC degradation, 63
Seed effects, lead, 80
Soil adsorption, carbaryl (table), 100, 101
Soil arsenic levels, Chilean cities (illus.), 128
Soil insecticide residues, Egypt (table), 31
Spodoptera littoralis resistance, insecticides (table), 11
Surface water, arsenic-removal process (illus.), 133
Surfactant biodegradation, PFCs, 61

**T**

Thermodynamic properties, fluorinated propanes (table), 65
Thermodynamics, organohalogens, 58

Toxaphene, Egyptian insect resistance, 11
Toxicity of carbaryl, acute (table), 109
Toxicity of carbaryl, animals, 108
Toxicity of carbaryl, aquatic animals (table), 108
Toxicity of carbaryl, humans, 112
Toxicity of carbaryl, subacute and chronic, 109, 110
Toxicity of lead, plants (illus.), 79
Toxicity of lead, plants, 77
Toxicity to organisms, carbaryl, 107
Toxicokinetics of carbaryl, mammals, 114
Toxicology, carbaryl, 95 ff.

**V**

Vegetable insecticide residues, Egypt (table), 31
Vegetable pesticide residues, Egypt (table), 29, 30
Vegetable residues, pesticides (table), 39
Vegetables, pesticide intake risks (table), 40
Vegetables, pesticide residue intake (table), 38
Vitamin role, reductive dechlorination, 60

**W**

Water contamination, pesticide residues, 12, 13
Water pollution in Chile, arsenic removal, 132
Water profile, Chile (table), 133
Water regulations for arsenic, Chile (table), 142
Water residues, carbaryl (table), 99
Water treatment in Chile, arsenic- removal facilities (table), 134

**X**

Xenobiotic residues, birds and plants (table), 24

MIX
Papier aus verantwortungsvollen Quellen
Paper from responsible sources
FSC® C105338

If you have any concerns about our products,
you can contact us on
**ProductSafety@springernature.com**

In case Publisher is established outside the EU,
the EU authorized representative is:
**Springer Nature Customer Service Center GmbH
Europaplatz 3, 69115 Heidelberg, Germany**

Printed by Libri Plureos GmbH
in Hamburg, Germany